SOTHIS PRESS

Sothis Press aims to completely reimagine the public domain book with an emphasis on universal access to captivatingly designed print editions. We believe that technology will enable every single book ever published to be perpetually accessible in beautiful print form. Sothis is beauty in ubiquity.

www.sothis.press
1 Waterhouse Square, London, ECIN 2ST

A Practical Handbook on the Distillation of Alcohol From Farm Products

Including the Processes of Malting; Mashing and Mascerating; Fermenting and Distilling Alcohol From Grain, Beets, Potatoes, Molasses, Etc;, With Chapters on Alcoholometry and the De-Naturing of Alcohol, for Use in Farm Engines, Automobiles, Launch Motors

F. B. Wright

SOTHIS PRESS

ISBN 978-1-0368-0297-4

A Practical Handbook on the Distillation of Alcohol From Farm Products is in the public domain.

This edition is published by Sothis Press The company is registered in England and Wales. The Company Number is 14900389.

A PRACTICAL HANDBOOK

ON THE

DISTILLATION OF ALCOHOL

FROM

FARM PRODUCTS

INCLUDING

The Processes of Malting; Mashing and Mascerating; Fermenting and Distilling Alcohol from Grain, Beets, Potatoes, Molasses, etc., with Chapters on Alcoholometry and the

DE-NATURING OF ALCOHOL

FOR USE IN

Farm Engines, Automobiles, Launch Motors, and in Heating and Lighting; with a Synopsis of the New Free Alcohol Law and its Amendment and the Government Regulations.

BY F. B. WRIGHT.

SECOND EDITION, REVISED AND GREATLY ENLARGED

NEW YORK
SPON & CHAMBERLAIN, 120 LIBERTY STREET,

LONDON
E. & F. N. SPON, LIMITED, 57 HAYMARKET, S.W.

1918

CAMELOT PRESS, 226 William Street, New York, U. S. A.

PREFACE TO SECOND EDITION.

Since the passage of the " Free Alcohol Act " there has been a constantly increasing demand for information as to the manufacture of industrial alcohol. This, with the favorable reception accorded to the first edition of this book has lead the publishers to bring out a second edition.

The entire volume has been carefully revised and not only has the original text been amplified but new chapters have been added explaining the most modern and approved methods and appliances both as used in Europe and in this country. Another valuable feature of the present volume is the collection of U. S. de-naturing formulas covering the special denaturants necessitated by the various arts and by the Government requirements. The chapters on modern distilling apparatus rectifiers and modern plants have been very carefully prepared in order to give the reader a clear idea of the various types of apparatus in use to-day and of their general place in a distillery system. The value of the book has been further increased by numerous additional illustrations.

It would be impossible in the compass of one small volume to describe all the practical details of alcohol manufacture particularly as these details vary with every distillery, but it has been the aim of the author to give sufficient information to enable every reader to understand the theory and general practice of the art, leading him from the simple methods and apparatus used until the last ten years to the more complicated stills and processes which have been lately devised.

Inasmuch as the manufacture of industrial alcohol has been most highly perfected in France and Germany, use has been made of the best European authorities and in particular the author begs to acknowledge his indebtedness to Sa Majeste L' Alcohol by L. Beaudry de Saunier. The publishers' and author's acknowledgements are also due to the Vulcan Copper Works Company of Cincinnati, Ohio, and to the Geo. L. Squier Manufacturing Company, Buffalo, New York, for their kindness in allowing illustrations to be given of modern American distilling apparatus.

F. B. Wright.

PREFACE.

To the majority of persons Alcohol connotes liquor. That it is used to some extent in the arts, that it is a fuel, is also common knowledge, but Alcohol as a source of power, as a substitute for gasoline, petroleum, and kindred hydrocarbons was hardly known to the generality of Americans until the passage of the " De-naturing Act " by the last Congress.

Then Alcohol leaped at once into fame,—not merely as the humble servant of the pocket lamp, nor as the Demon Rum, but as a substitute for all the various forms of cheap hydrocarbon fuels, and as a new farm product, a new means for turning the farmer's grain, fruit, potatoes, etc., into that greatest of all Powers, *Money*.

That Alcohol was capable of this work was no new discovery accomplished by the fiat of Congress, but the Act of June 7, 1906, freed de-natured Alcohol from the disability it had previously labored under,—namely, the high internal revenue tax, and so cheapened its cost that it could be economically used for purposes in the arts and manufactures which the former tax forbade.

This Act then opens the door of a new market to the farmer and the manufacturer, and it is in answer to the increased desire for information as to the source of Alcohol and its preparation that this book has been written. The processes described are thoroughly reliable and are such as have the approval of experience.

As was stated above, Alcohol is not a natural product, but is formed by the decomposition of sugar or glucose through fermentation. This leaves Alcohol mixed with water, and these in turn are separated by distillation.

The literature treating of the distillation of Alcohol from farm products is very scant. But due credit is here given to the following foreign works which have been referred to: Spon's Encyclopædia of the Industrial Arts, which also contains an article on Wood Alcohol, Mr. Bayley's excellent Pocketbook for Chemists, and Mr. Noel Deerr's fine work on Sugar and Sugar Cane.

NEW YORK.

CONTENTS.

CHAPTER IV.

MODERN DISTILLING APPARATUS.

CHAPTER V.

RECTIFICATION.

CHAPTER VI.

MALTING.

CHAPTER VII.

ALCOHOL FROM POTATOES.

CHAPTER VIII.

Alcohol from Grain, Corn, Wheat, Rice, and Other Cereals.

CHAPTER IX.

Alcohol from Beets.

CHAPTER X.

Alcohol from Molasses and Sugar Cane.

CHAPTER XI.

Alcoholometry.

LIST OF ILLUSTRATIONS.

CHAPTER I.

ALCOHOL, ITS VARIOUS FORMS AND SOURCES.

Alcohol. (Fr., *alcool*; Ger., *alkohol.*) Formula, C_2H_6O.

Pure alcohol is a liquid substance, composed of carbon, hydrogen, and oxygen, in the following proportions:

C..........................	52.17
H..........................	13.04
O..........................	34.79
	100.00

It is the most important member of an important series of organic compounds, all of which resemble each other closely, and possess many analogous properties. They are classed by the chemist under the generic title of " Alcohols."

Alcohol does not occur in nature; it is the product of the decomposition of sugar, or, more properly, of *glucose*, which, under the influence of certain organic, nitrogenous substances, called *ferments* is split up into alcohol and carbonic anhydride. The latter is evolved in the form of gas, alcohol remaining behind mixed with water, from which it is separated by distillation. The necessary purification is effected in a variety of ways.

Pure, absolute alcohol is a colorless, mobile, very volatile liquid, having a hot, burning taste, and a pungent and somewhat agreeable odor. It is very inflammable, burning in the air with a bluish-

TABLE I.—THE BOILING POINTS OF ALCOHOLIC LIQUORS OF DIFFERENT STRENGTHS, AND THE PROPORTIONS OF ALCOHOL IN THE VAPORS GIVEN OFF.

Proportion of alcohol in the boiling liquid in 100 vols.	Temperature of the boiling liquid.	Proportion of alcohol in the condensed vapor in 100 vols.	Proportion of alcohol in the boiling liquid in 100 vols.	Temperature of the boiling liquid.	Proportion of alcohol in the condensed vapor in 100 vols.
92	171.0 F.	93	20	189.5 F.	71
90	171.5 F.	92	18	191.6 F.	68
85	172.0 F.	91.5	15	194.0 F.	66
80	172.7 F.	90.5	12	196.1 F.	61
75	173.6 F.	90	10	198.5 F.	55
70	175.0 F.	89	7	200.6 F.	50
65	176 0 F.	87	5	203.0 F.	42
50	178.1 F.	85	3	205.1 F.	36
40	180.5 F.	82	2	207.5 F.	28
35	182.6 F.	80	1	209.9 F.	13
30	185.0 F.	78	0	212.0 F.	0
25	187.1 F.	76			

yellow flame, evolving much heat, leaving no residue, and forming vapors of carbonic anhydride and water. Its specific gravity at 0° C (32° F.) is .8095, and at 15.5° C. (60° F.) .794; that of its vapor is 1.613. It boils at 78.4° C. (173° F.). The boiling point of its aqueous mixtures are raised in proportion to the quantity of water present. Mixtures of alcohol and water when boiled give off at first a vapor rich in alcohol, and containing but little aqueous vapor; if the ebullition be continued a point is ultimately reached when all the alcohol has been driven off and nothing but pure water remains. Thus, by repeated distillations alcohol may be obtained from its mixtures with water in an almost anhydrous state.

Absolute alcohol has a strong affinity for water. It absorbes moisture from the air rapidly, and thereby becomes gradually weaker; it should therefore be kept in tightly-stoppered bottles. When brought into contact with animal tissues, it deprives them of the water necessary for their constitution, and acts in this way as an energetic poison. Considerable heat is disengaged when alcohol and water are brought together; if, however, ice be substituted for water, heat is absorbed, owing to the immediate and rapid conversion of the ice into the liquid state. When one part of snow is mixed with two parts of alcohol, a temperature as low as 5.8° F. below zero is reached.

When alcohol and water are mixed together the resulting liquid occupies, after agitation, a less volume than the sum of the two original liquids. This contraction is greatest when the mixture is made in the proportion of 52.3 volumes of alcohol and 47.7 volumes of water, the result being, instead of 100 volumes, 96.35. A careful examination of the liquid when it is being agitated reveals a vast number of minute air-bubbles, which are discharged from every point of the mixture. This is due to the fact that gases which are held in solution by the alcohol and water separately are less soluble when the two are brought together; and the contraction described above is the natural result of the disengagement of such dissolved gases. The following table represents the contraction undergone by different mixtures of absolute alcohol and water.

TABLE II.—100 VOLUMES OF MIXTURE AT 59° F.

Alcohol.	Contraction.	Alcohol.	Contraction.	Alcohol.	Contraction.
100	0.00	65	3.61	30	2.72
95	1.18	60	3.73	25	2.24
90	1.94	55	3.77	20	1.72
85	2.47	50	3.74	15	1.20
80	2.87	45	3.64	10	0.72
75	3.19	40	3.44	5	0.31
70	3.44	35	3.14		

Alcohol is termed " absolute " when it has been deprived of every trace of water, and when its composition is exactly expressed by its chemical formula. To obtain it in this state it must be subjected to a series of delicate operations in the laboratory, which it would be impossible to perform on an industrial scale. In commerce it is known only in a state of greater or less dilution.

Alcohol possesses the power of dissolving a large number of substances insoluble in water and acids, such as many inorganic salts, phosphorus, sulphur, iodine, resins, essential oils, fats, coloring matters, etc. It precipitates albumen, gelatine, starch, gum, and other substances from their solutions. These properties render it an invaluable agent in the hands of the chemist.

Alcohol is found in, and may be obtained from, all substances—vegetable or other—which contain sugar. As stated above, it does not exist in these in the natural state, but is the product of the decomposition by fermentation of the saccharine principle contained therein; this decomposition yields the spirit in a very dilute state, but it is readily separated from the water with which it is mixed by processes of distillation, which will subsequently be described. The amount of alcohol which may be obtained from the different unfermented substances which yield it varies considerably, depending entirely upon the quantity of sugar which they contain.

Alcohol is produced either from raw materials containing starch, as potatoes, corn, barley, etc., or raw materials containing sugar, as grapes, beets, sugar-cane, etc.

The following are some of the most important sources from which alcohol is obtained: Grapes, apricots, cherries, peaches, currents, gooseberries, raspberries, strawberries, figs, plums, bananas, and many tropical fruits, artichokes, potatoes, carrots, turnips, beet-root, sweet corn, rice and other grains. Sugar-cane refuse, sorgum, molasses, wood, paper, and by a new French process from acetylene. On a large scale alcohol is usually obtained from sugar beets, molasses or the starch contained in potatoes, corn and other grains. The starch is converted into maltose by mixing with an infusion of malt. The maltose is then fermented by yeast. Sulphuric acid may be used to convert even woody fibre,

TABLE III.—PRINCIPAL ALCOHOLS.

Chemical Name.	Source.	Formula.	Boiling Point °F.
1 Methyl Alcohol	Distillation of Wood	CH_3OH	150.8
2 Ethyl "	Fermentation of sugar	C_2H_5OH	172.4
3 Propyl "	" " grapes	C_3H_7OH	206.6
4 Butyl "	" " beets	C_4H_9OH	242.6
5 Amyl "	" " potatoes	$C_5H_{11}OH$	278.6
6 Caproyl "	" " grapes	$C_6H_{13}OH$	314.6
7 Aenanthyl "	Distillation castor oil with potatoes	$C_7H_{15}OH$	347.
8 Capryl "	Essential oil hog weed	$C_8H_{17}OH$	375.8
9 Nonyl "	Nonane from petroleum	$C_9H_{19}OH$	
10 Rutyl "	Oil of Rue	$C_{10}H_{21}OH$	
11 Cytyl "	Spermaceti	$C_{16}H_{33}OH$	
12 Ceryl "	Chinese wax	$C_{26}H_{53}OH$	
13 Melisyl "	Bees' wax	$C_{30}H_{61}OH$	

paper, linen, etc., into glucose, which may in turn be converted into alcohol.

Among a variety of other substances which have been and are still used for the production of alcohol in smaller quantities, are roots of many kinds, such as those of asphodel, madder, etc. Seeds and nuts have been made to yield it. It will thus be seen that the sources of this substance are practically innumerable; anything, in fact, which contains or can be converted into sugar is what is termed " alcoholisable."

Alcohol has become a substance of such prime necessity in the arts and manufactures, and in one form or another enter so largely into the composition of the common beverages consumed by all classes of people that its manufacture must, of necessity, rank among the most important industries of this and other lands.

Of the alcohols given in the above table only two concern the ordinary distiller, or producer of alcohol for general use in the arts. Methyl alcohol, the ordinary " wood alcohol," or wood naphtha, and Ethyl alcohol, which is produced by the fermentation of sugar and may therefore be made from anything which contains sugar.

Ethyl alcohol forms the subject of this treatise. Aside from its chemical use in the arts as a source of energy and as a fuel, alcohol will likely soon compete with petroleum, gasoline, kerosene, etc., under the Act of Congress freeing the "de-naturized" spirit from the Internal Revenue tax. This act and the de-naturing process are covered in the last chapters of this book.

CHAPTER II.

The Preparation of Mashes, and Fermentation.

Alcohol may be produced either from, (1) farinacious materials, such as potatoes or grains, (2), from sacchariferous substances such as grapes, sugar beets, sugar cane, or the molasses produced in sugar manufacture.

THE PREPARATION OF STARCHY MATERIALS.

Saccharification. Preparatory Mashing. With starchy materials it is first necessary to convert the starch into a sugar from which alcohol can be produced by the process of fermentation. This is called saccharification.

Gelatinizing. The first step in this process is gelatinizing the starch;—that is, forming it into a paste by heating it with water, or into a liquid mass by steaming it under high pressure. The liquid or semi-liquid mass is then run into a preparatory mash vat and cooled.

Saccharifying. The disintegrated raw materials or gelatinized starch in the preparatory mash vat is now to be "saccharified" or converted into sugar.

This is effected by allowing malt to act on the starch. This malt contains a certain chemical "ferment" or enzyme, called "diastase" ("I separate").

This is able under proper conditions to break up the gelatinized starch into simpler substances—the dextrins—and later into a fermentable sugar called maltose.

Fermentation.—The maltose or sugar in the "mash" is now to be converted into alcohol. This is accomplished by fermentation, a process of decomposition which converts the sugar into carbonic acid and alcohol. Fermentation is started by yeast, a fungus growth, which in the course of its life history produces a matter called zymose which chemically acts on the sugar to split it up into carbonic acid gas and alcohol.

Yeast may be either "wild" or cultivated. If the mash is left to stand under proper condition the wild yeast spores in the air, will soon settle in the mash and begin to multiply. This method of fermentation is bad because other organisms than yeast will also be developed,—organisms antagonistic to proper fermentation. As a consequence, pure or cultivated yeast is alone used.

This yeast is cultivated from a mother bed in a special yeast mash and when ripened is mixed with the mash in the fermenting vat. At a temperature between 50° F. and 86° F. the yeast induces fermentation, converting the sugar of the mash into carbon dioxid which escapes, and alcohol which remains in the decomposed mash, or " beer " as it is termed in the United States.

It now remains to separate the alcohol from the water of the beer with which it is mixed. This is accomplished by distillation and rectification, as will be fully described in the chapters following.

PRODUCTION OF ALCOHOL FROM SACCHARIFEROUS SUBSTANCES.

Substances such as grape juice, fruit juice, sugar beets, cane sugar and molasses already contain fermentable sugar. Saccharification is therefore not needed and juices or liquids from these matters are either directly fermented as in the case of sugar cane, or—as in the case of sugar beets—the sugar in juice is transferred by yeast into a fermentable sugar.

MASHING STARCHY MATERIALS.

We will now consider in more detail the preparation of mashes from starch-containing substances.

Gelatinizing Apparatus. These comprise either ordinary vats, into which steam at low pressure is admitted (see Fig. 44), cookers and stirrers such as shown in Fig. 1 and 45 or the Henze steamer (Fig. 2.)

An example of a cooking and mashing apparatus and its connections is shown in Fig. 1. This is the vacuum cooker put on the market by the

Vulcan Copper Works Company, of Cincinnati Ohio. This consists of a cylindrical steel vessel the interior of which is fitted with stirrer arms attached to a shaft making about sixty revolutions per minute. The steam enters the vessel at the bottom by means of pipes conducting it from a manifold, or header, in the same manner as is shown in the apparatus illustrated in Fig 45. Attached to each pipe at its point of entrance is a check valve to spray the steam through the mash. A thermometer for registering the temperature and a water gauge are placed in the manifold. The grain enters the cooker from the grain hopper by way of a spout. The cylinder has been previously supplied with hot water and during the mixing of the meal with the hot water the mass is constantly stirred. The malt is mixed with water in the small grain tub which is provided with a stirrer. The malt mash is admitted into the cooker and the mass thoroughly mixed by the arms. After the mashing, the product passes off to the drop tub and from thence to the mash coolers where it is cooled to the proper temperature for fermentation. The gearing for agitating the malt mash and the grain or potato mash is evident from the drawing.

The pressure steamers used in mashing are shown in Fig. 2. They comprise a cylindrical vessel preferably conical or partly conical, provided with steam entrance pipes, air valves and a manhole. At the bottom of the cone forming the lower end of the steamer is a grating located in an exit pipe provided with a valve. One of the steam entrance pipes is so located that the steam is forced in at the top of the cylinder while the

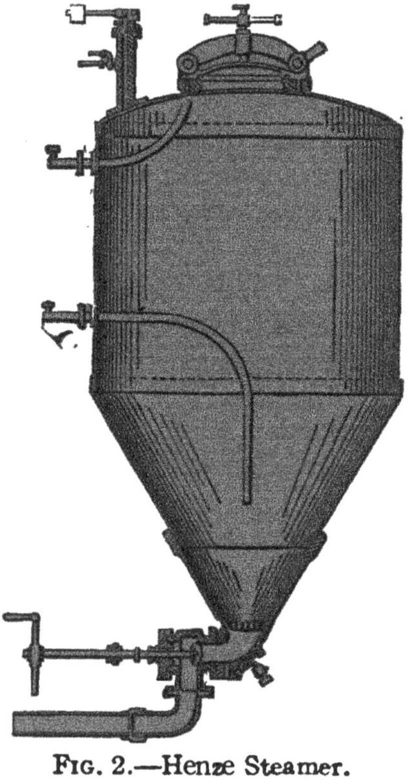

FIG. 2.—Henze Steamer.

other allows steam to enter at the bottom of the cylinder. The device is provided with a pressure gauge and an air cock.

In use the body of the apparatus is partly filled

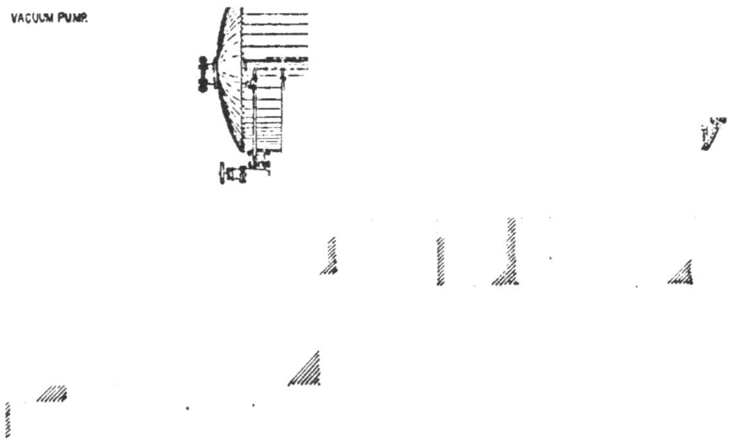

FIG. 1.—Vacuum Mash Cooker.

(To face page 10)

with water and the material to be treated. This is acted upon by a steam pressure of two atmospheres, which is later increased to three, steam entering by the lowermost pipe, passing up through the water and potatoes thoroughly agitating the same and passing away by the steam gauge. After standing at the last pressure for ten or fifteen minutes the lower steam inlet is closed; the upper inlet and the blow-out valve are opened. The steam is then increased to its highest point or about four atmospheres and the lower valve is opened. The disintegrated material is forced out by the steam through the grating at the bottom of the cone. This comminutes it and pulps it before it passes into the preparatory mash tub. Blowing out requires about 40 to 50 minutes. Steaming and blowing out together cover a space of two hours. The pressure of the steam before blowing out should be such that the steam is constantly being blown off through the safety valve. Thus the mass in the steamer is agitated and the material entirely disintegrated and gelatinized.

Process. Into these apparatuses the potatoes and corn or grain first ground into mash, or even corn or grain unground, if the pressure is high enough, are disintegrated and cooked by steam under high pressure. During this process the starch becomes partially dissolved and partially gelatinized, which occurs when a pressure of some 65 pounds has been attained, with a temperature of about 300° F.

Saccharifying. It is now necessary to saccharify the gelatinized mass. This is accomplished by adding to it a certain amount of malt, whereby maltose or sugar is formed through the action of the diastase. The amount of maltose so created is in proportion to the amount of malt used, the length of time it is acting, the dilution of the mash, and the existence of a proper temperature. The temperature best fitted for this action lies above 122° F., but in order to entirely dissolve the starch a temperature of 145° F. should be used. In addition, at this higher temperature, the bacteria inimical to fermentation are destroyed. A higher temperature than 145° F. should not be allowed. except in extraordinary cases as it injures the effectiveness of the diastase.

Apparatus. The mixture of the malt with the mash may either take place in the heater and cooker itself (see Fig. 2) or in a preparatory mash vat.

In the first instance, the malt is allowed to enter the cooking cylinder when the temperature of the mash is about 145° F. The mash is stirred until thoroughly mixed when the product is drawn into a receptacle called a drop tub and later reduced to a proper fermenting temperature.

When the Henze type of steamer is used, the pulped mass (see Page 121) is blown into a preparatory mash vat, at the proper temperature. It is left to stand at this temperature for a period varying from twenty minutes to an hour and a half.

Cooling the Mash. Saccharification takes place at a temperature above 122° F., but the proper fermenting temperature is only about 63° F. to 68° F., and hence some means must be adopted for cooling the hot mash to this tempreature and for so cooling it in a relatively short time.

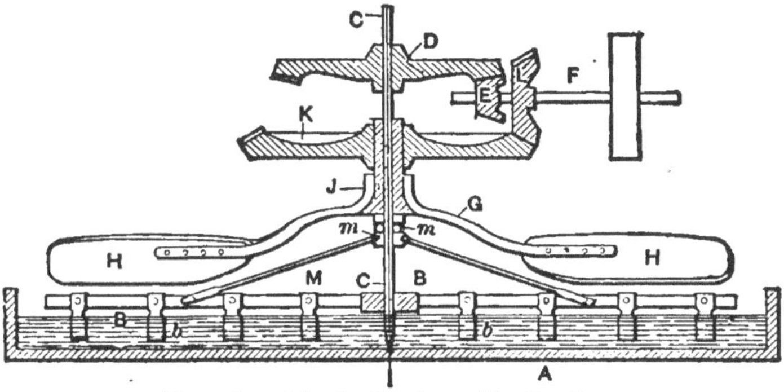

Fig. 3.—Mash Cooler, Air System.

Cooling may be accomplished by submitting the mash to currents of air; to contact with cold water coils or by the use of ice. One of the simplest coolers of the first class is shown in Fig. 3.

This consists of a shallow panlike tank *A* having means for introducing and drawing off the mash. Rotating in the center of the tank is a vertical shaft *C* carrying radiating stirrer arms B. Braces *M* extend to the middle of these arms and the arms carry a number of blades or paddles *b*, which extend down into the mash. Above the arms, mounted loosely on the same shaft, but rotating in the opposite direction, are fans *H* supported by arms *J* which create air currents over the agitated mash. These fans move at a much faster rate than the stirrers *B*.

A simple form of driving gear is shown. The main shaft *C* is rotated by a large bevel gear *D*, meshing with a small pinion *E* on the end of a driving shaft F, which is driven by a belt. This shaft also carries a bevel gear *L*, which meshes with a bevel gear *K* mounted on a sleeve. This sleeve surrounds and rotates freely on the central shaft *C*, being supported at its lower end in ball bearings *m m*, mounted on the shaft. This combination gives opposite rotation to the faces and stirrer arms and at different speeds. The driving mechanism can be of course varied.

Another simple method of air cooling would be to let the mash run down a series of enclosed steps or chutes, the casing being kept cool by an air blast. Mashes may be even cooled by mere stirring by paddles, but this takes a long time and much labor.

The preparatory mash vats used to-day are almost all provided with stirrers formed of hollow blades capable of a rapid stirring movement through the mash. Through the hollow blades cold water is forced. Mash vats of this kind should have the following qualities. They should be strongly built, particularly as regards the stirrers so as to be used with thick mashes. They should thoroughly and uniformly stir and mix the mash and they should be capable of cooling the mash within an hour, and should be so constructed as to be easily cleaned.

By using coils of pipe which may be inserted or withdrawn from the mash tub, and through which cold water is forced, the mash may be effectively cooled, but the best plan for quick cooling is to bring

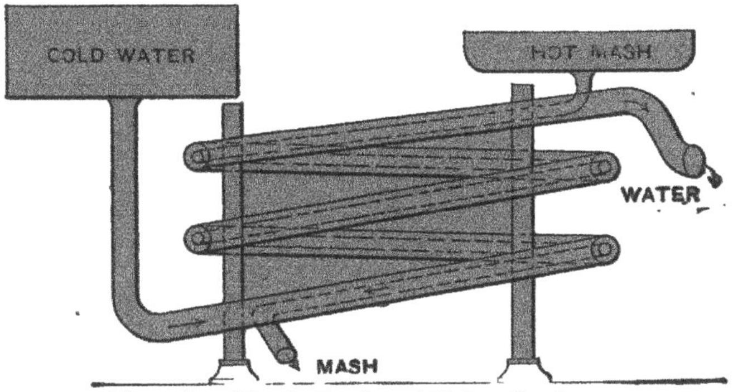

FIG. 4.—Mash Cooler, Water System.

a comparatively thin layer of the mash in contact with the coils. This may be conveniently done by using a system of comparatively large water pipes enclosing small pipes for the passage of the mash.

This should be arranged in a stand like the coils of a radiator with an incline from the inlet end of the top pipe to the outlet end of the lowermost pipe. As stated, the small pipe carries the mash, the large pipe the water.

Preferably the mash flows downward while the water is forced upward in a contrary direction by means of a pump or a high level reservoir. The cooled mash should flow into the fermenting tank at a temperature of about 68° F.

There are many varieties of mash cooling apparatuses on the market of more or less complication suited to the needs of large and expensive plants,

The form of cooler best to be used depends upon the circumstances of each case and whether thick or thin mashes are to be distilled. The cooler should, however, be capable of thorough cleansing so that no portion of one mashing be carried to another.

Fermentation is an obscure and seemingly spontaneous change or decomposi.ion which takes place in most vegetable and animal substances when exposed at ordinary temperatures to air and moisture. While fermentation broadly covers decay or putrifaction, yet it is limited in ordinary use to the process for producing alcoholic liquors from sacchariferous mashes.

Fermentation is brought about by certain bodies called ferments—these are either organized, as vegetable ferments such as yeast, or unorganized as diastase,—the enzyme of germinated malt. The last is used to convert starch into maltose, the first is used to convert maltose into fermentable sugar. The organized ferments are either to be found floating freely in the air under the name of wild yeast or are artificially produced. If a solution of pure sugar be allowed to stand so that it can be acted on by the organisms in the air, it will remain unaltered for a long time, but finally mold will appear upon it and it will become sour and dark-colored. If, however, a suitable ferment is added to it, such as yeast, it rapidly passes into a state of active fermentation by which the sugar is split up into alcohol and carbon dioxid, the process continuing from 48 hours to several weeks according to the temperature, the amount of sugar present, and the nature and quantity of the ferment. Fermentation cannot occur at a temperature much below 40° F., nor above 140° F. The limits of practical temperature, however, are 41° to 86° F. Brewer's yeast is chiefly employed in spirit manufacture.

The most striking phenomena of fermentation are the turbidity of the liquid, the rising of gas bubbles to the surface, and the increase in temperature, the disappearance of the sugar, the appearance of alcohol and the clearing of the liquid. At the end a slight scum is formed on the top of the liquid and a light colored deposit at the bottom. This deposit consists of yeast which is capable of exciting the vinous fermentation in other solutions of sugar. The lower the temperature the slower the process, while at a temperature above 86° F. the vinous fermentation is liable to pass into other forms of fermentation to be hereafter considered.

There are many theories of fermentation, of which the two most important are those of Pasteur and Buchner. The first teaches that fermentation is caused purely by the organic life of the yeast plant and is not a mere chemical action, whereas the second view most largely held to-day is that fermentation is a purely chemical change due to certain unorganized substances called "enzymes" present in the yeast.

The theory need not detain us. It is sufficient that the yeast plant in some manner acts to decompose the saccharified mash into alcohol and carbonic acid gas.

Yeast is a fungus, a mono-cellular organism, which under proper conditions propogates itself to an enormous extent. There are many races or varieties of yeast each having its peculiar method of growth.

For our purposes we may divide the yeast races into two classes, wild yeast and cultivated yeast. Originally any of the yeast races were supposed to be good enough to effect fermentation but to-day every effort is made to procure and use only those races which have the greatest power to decompose sugar. It was for this reason that the old distiller kept portions of his yeast over from one fermentation to the next. This was yeast whose action they understood and whose abilities were proven. This yeast so kept was open, however, to the chance of contamination and yeast to-day is as carefully selected and bred as is a strain of horses, or dogs, or plants.

After getting a portion of selected pure yeast for breeding purposes, it may be sowed, that is,

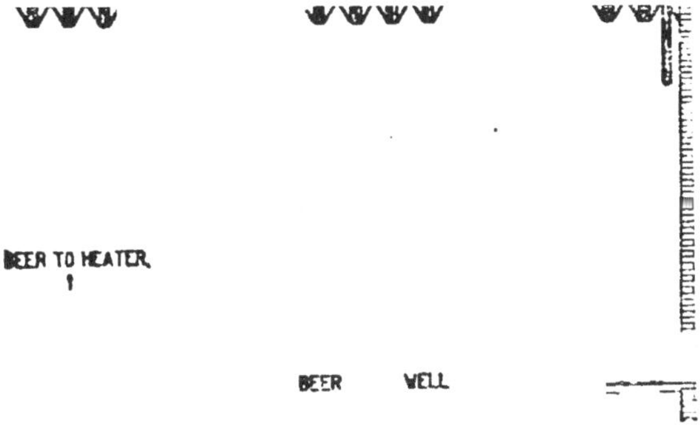

Fig. 5.—Yeasting and Ferme

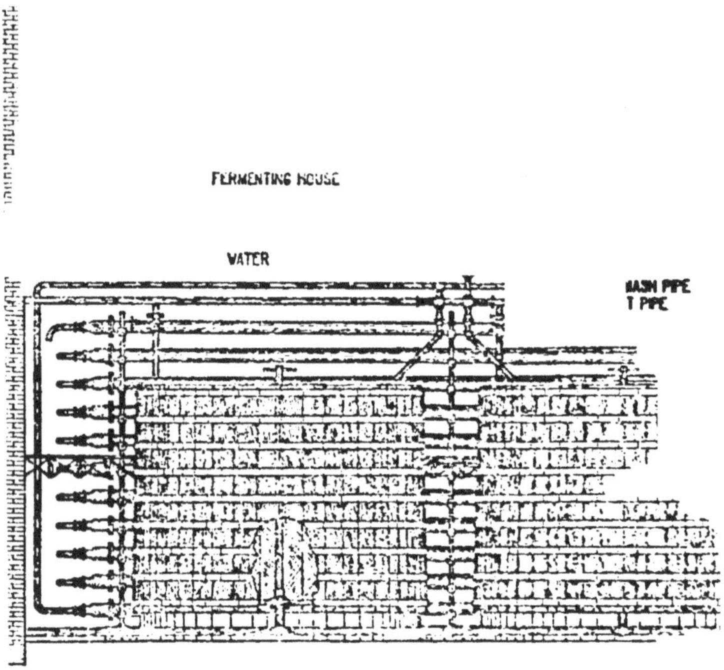

nting Apparatus.

(To face pag

propagated very carefully in a yeast mash, in sterilizing apparatus, where all chance of contamination by bacteria or wild yeast is avoided. From this bed of mother yeast, or start yeast, the yeast for the successive yeast mashes is taken.

The preparation of the various varieties of yeast mashes is too lengthy to be set forth except in special treatises on the subject, but the ordinary method of yeasting is as follows, reference being made to Fig. 5, which shows the apparatus used in the yeasting and fermenting departments of a distillery, as installed by the Vulcan Copper Works, of Cincinnati. The yeast tubs are shown to the left of the illustration. They are each provided with cooling coils and stirrers.

The yeast mash we will assume is composed of equal parts of barley malt and rye meal. Hot water at 166° F. is first put into the mash tub. The rake or stirrers are then rotated and the meal run in slowly. The stirring is continued for twenty minutes after the meal is all in, during which the mash has become saccharified.

The mash is then allowed to stand for about twenty hours, and to grow sour by lactic fermentation. The lactic acid so produced protects the mother yeast from infection by suppressing wild yeast and bacteria. During this period great care is taken to prevent the temperature of the mash falling below 95° F. and consequent butyric and acetous fermentation following. After it has so stood the sour mash is cooled by circulating water in the coils and stirring until it is reduced to from 59° to 68° F. depending on whether the mash is thin or thick. Start yeast during the cooling of the mash when at above 86° F. is added and stirred in. For the next twelve hours the yeast ferments and when a temperature of 84° F. has been attained the mash is cooled to 65° F. at which temperature it is maintained until allowed to enter the fermenting tubs through the pipe leading thereto from the yeast tub.

There are four principal kinds of fermentation: alcoholic, acetous, lactic and viscous.

Alcoholic Fermentation. This may be briefly described as follows: The mash in the fermenting vat having been brought to the proper temperature, the ferment is thrown in, and the whole is well stirred together.

This is know as pitching.

The proper pitching temperature varies with the method of fermentation adopted, the length of the fermenting period, the materials of the mash, its thickness or attenuation. It must always be remembered that there is a great increase in the temperature of the "beer" during fermentation and that the temperature at its highest should never under any circumstances, become greater than 86° F. and with thick mashes that even a less heat is desirable. Therefore the pitching temperature should be such that the inevitable rise due to fermentation shall not carry the temperature to or beyond the maximum point desired for the particular mash being treated. It is to accurately control the pitching temperature and the fermenting temperature that the fermenting tanks are provided with cooling appliances.

In about three hours' time, the commencement of the fermentation is announced by small bubbles of gas which appear on the surface of the vat, and collect around the edges. As these increase in number, the whole contents are gradually thrown into a state of motion, resembling violent ebullition, by the tumultuous disengagement of carbonic anhydride. The liquor rises in temperature and becomes covered with froth. At this point, the vat must be covered tightly, the excess of gas finding an exit through holes in the lid; care must now be taken to prevent the temperature from rising too high, and also to prevent the action from becoming too energetic, thereby causing the contents of the vat to overflow. In about twenty-four hours the action begins to subside, and the temperature falls to that of the surrounding atmosphere. An hour or two later, the process is complete; the bubbles disappear, and the liquor, which now possesses the characteristic odor and taste of alcohol, settles out perfectly clear. The whole operation, as here described, usually occupies from forty-eight to seventy-two hours. The duration of the process is influenced, of course, by many circumstances, chiefly by the bulk of the liquor, its richness in sugar, the quality of the ferment, and the temperature.

Acetous Fermentation. This perplexing occurrence cannot be too carefully guarded against. It results when the fermenting liquor is exposed to the air. When this is the case, the liquor absorbs a portion of the oxygen, which unites with the alcohol, thus converting it into acetic acid as rapidly as it is formed. When acetous fermentation begins, the liquor becomes turbid, and a long, stringy substance appears, which after a time settles down to the bottom of the vat. It is then found that all the alcohol has been decomposed, and that an equivalent quantity of acetous acid remains instead. It has been discovered that the presence of a ferment and a temperature of 68° to 95° F. are indispensable to acetous fermentation, as well as contact with the atmosphere. Hence, in order to prevent its occurrence, it is necessary not only to exclude the air, but also to guard against too high a temperature and the use of too much ferment. The latter invariably tends to excite acetous fermentation. It should also be remarked that it is well to cleanse the vats and utensils carefully with lime water before using, in order to neutralize any acid which they may contain; for the least trace of acid in the vat has a tendency to accelerate the conversion of alcohol into vinegar. A variety of other circumstances are favorable to acetification, such as the use of a stagnant or impure water, and the foul odors which arise from the vats; stormy weather or thunder will also engender it.

Lactic Fermentation. Under the influence of lactic fermentation, sugar and starch are converted into lactic acid. When it has once begun, it develops rapidly, and soon decomposes a large quantity of glucose, but as it can proceed only in a neutral liquor, the presence of the acid itself speedily checks its own formation. Then, however, another ferment is liable to act upon the lactic acid already formed, converting it into *butyric acid*, which is easily recognized by its odor of rank butter. Carbonic anhydride and hydrogen are evolved by this reaction. The latter gas acts powerfully upon glucose, converting it into a species of gum called *mannite*, so that lactic fermentation—in itself an intolerable nuisance—becomes the source of a new and equally objectionable waste of sugar. It can be avoided only by keeping the vats thoroughly clean; they should be washed with water acidulated with five per cent. of sulphuric acid. An altered ferment, or the use of too small a quantity, will tend to bring it about.

The best preventives are *thorough cleanliness*, and the use of good, fresh yeast in the correct proportion.

Viscous Fermentation. This is usually the result of allowing the vats to stand too long before fermentation begins. It is characterized by the formation of viscous or mucilaginous matters, which render the liquor turbid, and by the evolution of carbonic anhydride and hydrogen gases the latter acting as in the case of lactic fermentation and converting the glucose into mannite. Viscous fermentation may generally be attributed to the too feeble action of the ferment. It occurs principally in the fermentation of white wines, beer, and beet-juice, or of other liquors containing much nitrogenous matter. It may be avoided by the same precautions as are indicated for the prevention of lactic fermentation.

Periods of Fermentation. The operation of fermentation may be conveniently divided into three equal periods.

The first or pre-fermentation period is that when the yeast mixed into the mash is growing; the temperature should then be kept at about 63 to 68° F. during which time the yeast is propagated. The growth of the yeast is manifested by the development of carbonic acid gas and by a slight motion of the mash. When alcohol is produced to an extent of say five per cent. the growth of the yeast stops

The second period of chief fermentation then begins. Carbonic acid is freely developed and the sugar is converted into alcohol. The temperature at this time should not exceed 81.5° F. The second period of fermentation continues about 12 hours, when the last period commences.

During the third period or after fermentation there is a lessening of the formation of carbonic acid and a lowering of the temperature. In this stage the mash is kept at a temperature of 77° to 81° F.

In order to conveniently regulate the temperature of the mash the vat may be provided with a copper worm at the bottom thereof, through which cold water is forced. This, however, need only be used for thick mashes. There are also various kinds of movable coolers used for this purpose.

There are a number of different forms which fermentation may take. The insoluble constituents of the mash in the process of fermentation are forced to the surface, and form what may be termed a cover. If the carbonic acid gas bubbles seldom break this cover it indicates that the conversion of the sugar into alcohol and carbonic acid is proceeding very slowly and imperfectly. If, however, the cover is swirling and seething, and particularly if the cover is rising and falling with every now and then a discharge of gas, it is an indication that the conversion is properly proceeding. Foaming of the mash is to be prevented, as the froth or foam flows over the mash tank and considerable loss is sustained. It may be prevented by pouring a little hot lard into the vat, or petroleum, provided its odor will not interfere with the use of the alcohol when distilled.

Water is added in small quantities near the termination of the second period of fermentation. This dilutes the alcohol, in the mash and lessens its percentage, and thus the further growth of the yeast is permitted.

After fermentation the mash takes either the form of a thick diluted pulp or of a thin liquor. Again the reader is reminded that the mash after fermentation contains alcohol mixed with water—and that the next step in the process—distillation is necessary merely to separate the alcohol from the water.

There is always some loss in the process of fermentation; in other words, the actual production is below the theoretical amount due. Theoretically one pound of starch should yield 11.45 fluid ounces of alcohol. With a good result 88.3 per cent. of this theoretical yield is obtained; with an average result of 80.2 per cent. and with a bad result only about 72.6 per cent. or less.

Fermenting Apparatus. It remains now to describe briefly the vessels or vats employed in the processes of fermentation. They are made of oak or cypress, firmly bound together with iron bands, and they should be somewhat deeper than wide, and slightly conical, so as to present as small a surface as possible to the action of the air. Their dimensions vary, of course, with the nature and quantity of the liquor to be fermented. Circular vats are preferable to square ones, as being better adapted to retain the heat of their contents. The lid should close securely, and a portion of it should be made to open without uncovering the whole. For the purpose of heating or cooling the contents when necessary, it is of great advantage to have a copper coil at the bottom of the vat, connected with two pipes, one supplying steam and the other cold water.

Iron vats have also been used, having a jacketed space around them, into which hot or cold water may be introduced. As wooden vats are porous and hence uncleanly they have to be constantly scrubbed and disinfected. It is advisable to cover the interior with linseed oil, varnish or with a shellac varnish. The diameter of the coil varies according to the size of the vat.

The room in which the vats are placed should be made as free from draughts as possible by dispensing with superfluous doors and windows; it should not be too high and should be enclosed by thick walls in order to keep in the heat. As uniformity of temperature is highly desirable, a thermometer should be kept in the room, and there should be stoves for supplying heat in case it be required. The temperature should be kept between 64° F. and 68° F.

Every precaution must be taken to ensure the most absolute cleanliness; the floors should be swept or washed with water daily, and the vats, as pointed out above, must be cleaned out as soon as the contents are removed. For washing the vats, lime-water should be used when the fermentation has been too energetic or has shown a tendency to become acid; water acidulated with sulphuric acid is used when the action has been feeble and the fermented liquor contains a small quantity of undecomposed sugar. Care must be taken to get rid of carbonic anhydride formed during the operation. Buckets of lime-water are sometimes placed about the room for the purpose of absorbing this gas; but the best way of getting rid of it is to have a number of holes, three or four inches square, in the floor, through which the gas escapes by reason of its weight. The dangerous action of this gas and its effects upon animal life when unmixed with air are too well know to necessitate any further enforcement of these precautions.

The beer obtained by mashing and fermenting consist essentially of volatile substances, such as water, alcohol, essential oils and a little acetic acid, and of non-volatile substances, such as cellulose, dextrine, unaltered sugar and starch, mineral matters, lactic acid, etc.

The volatile constituents of the liquor possess widely different degrees of volatility; the alcohol has the lowest boiling point, water the next, then acetic acid, and last the essential oils. It will thus be seen that the separation of the volatile and non-volatile constituents by evaporation and condensation of the vapors given off is very easily effected, and that also by the same process, which is termed *distillation*, the volatile substances may be separated from one another. As the acetic acid and essential oils are present only in very small quantities, they will not require much consideration.

The aim of distillation is to separate as completely as possible the alcohol from the water which dilutes it. Table I shows the amount of alcohol contained in the vapors given off from alcoholic liquids of different strength, and also their boiling points.

A glance at this table shows to what an extent an alcoholic liquor may be strengthened by distillation, and how the quantity of spirit in the distillate increases in proportion as that contained in the original liquor diminishes. It will also be seen that successive distillations of spirituous liquors will ultimately yield a spirit of very high strength.

As an example, suppose that a liquid containing five per cent. of alcohol is to be distilled. Its vapor condensed gives a distillate containing 42 per cent. of alcohol which, if re-distilled, affords another containing 82 per cent. This, subjected again to distillation, yields alcohol of over 90 per cent. in strength. Thus three successive distillations have strengthened the liquor from five per cent. to 90 per cent.

It will thus be clear that the richness in alcohol of the vapors given off from boiling alcoholic liquids is not a constant quantity, but that it necessarily diminishes as the ebullition is continued. For example a liquor containing seven per cent. of alcohol yields, on boiling a vapor containing 50 per cent. The first portion of the distillate will, therefore, be of this strength. But as the vapor is proportionally richer in alcohol, the boiling liquor must become gradually weaker, and, in consequence, must yield weaker vapors. Thus, when the proportion of alcohol in the boiling liquid has sunk to five per cent., the vapors condensed at that time will contain only 40 per cent.; at two per cent. of alcohol in the liquor, the vapors yield only 28 per cent., and at one per cent., they will be found when condensed to contain only 13 per cent. From this it will be understood that if the distillation be stopped at any given point before the complete volatilization of all the alcohol the distillate obtained will be considerably stronger than if the process had been carried on to the end. Moreover, another advantage derived from checking the process before the end, and keeping the last portions of the distillate separate from the rest, besides that of obtaining a stronger spirit, is that a much purer one is obtained also. The volatile, essential oils, mentioned above, are soluble only in strong alcohol, and insoluble in its aqueous solutions. They distill also at a much higher temperature than alcohol, and so are found only among the last products of the distillation, which results from raising the temperature of the boiling liquid. This system of checking the distillation and removing the products at different points is frequently employed in the practice of rectification.

CHAPTER III.

DISTILLING APPARATUS.

The Apparatus employed in the process of distillation is called a *still*, and is of almost infinite variety. A still may be any vessel which will hold and permit fermentated "wash" or "beer" to be boiled therein, and which will collect the vapors arising from the surface of the boiling liquid and transmit them to a condenser. The still may be either heated by the direct application of fire, or the liquid in the still raised to the boiling point by the injection of steam. The steam or vapor rising from the boiling liquid must be cooled and condensed. This is done by leading it into tubes surrounded by cold water or the "cold mash."

The very simplest form of still is shown in Fig. 6, and consists of two essential parts, the still, or boiler *A*, made of tinned copper, the condenser *C* which may be made of metal or wood and the worm *B* made of a coil of tinned copper pipe.

The liquor is boiled in *A* and the vapors pass off into the worm *B*, which is surrounded by the cold water of the condenser, the distillate being drawn off at *f*.

The heated vapors passing through the worm *B* will soon heat up the water in *C* thereby retarding perfect condensation. To prevent this, a cold water supply pipe may be connected to the bottom of *C* making a connection at the top of *C* for an over flow of the warmed up water. By this means the lowest part of the worm will be kept sufficiently cool to make a rapid condensation of the vapors.

The boiler *A* can be made in two parts; the upper part fitting into the lower part snugly at

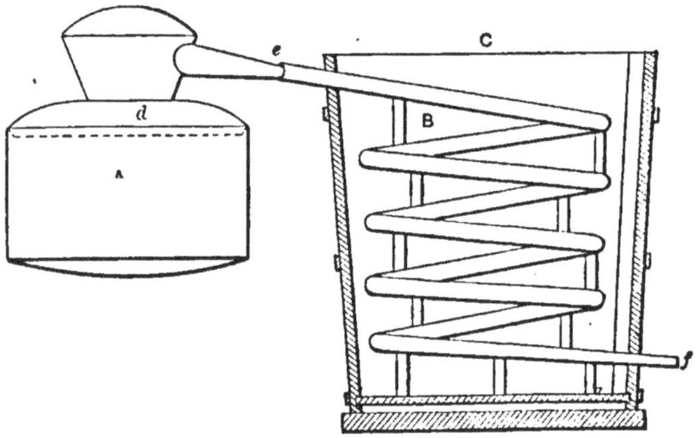

Fig. 6.—A Simple Still.

d. The pipe from the upper part fitting the worm snugly at *e*. This will enable the operator to thoroughly cleanse the boiler before putting in a new lot of liquor. The joints at *e* and *d* should be luted with dough formed by mixing the flour with a small portion of salt and moistening with water. This is thoroughly packed at the junctions of the parts to prevent the escape of steam or vapor.

Fig. 7 shows such a Still as manufactured by the Geo. L. Squier Mfg. Co., Buffalo, N. Y.

In an apparatus of this kind, the vapors of alcohol and water are condensed together. But if instead of filling the condenser *C* with cold water, it is kept at a temperature of 176° F. the greater part of the water-vapor will be condensed while the alcohol, which boils at 172.4° F. passes through the coil uncondensed. If therefore the

FIG. 7.—Simple Direct-Heated Still.

water be condensed and collected separately in this manner, and the alcoholic vapors be conducted into another cooler kept at temperature below 172.4° F., the alcohol will be obtained in a much higher state of concentration than it would be by a process of simple distillation

Supposing, again, that vapors containing but a small quantity of alcohol are brought into contact with an alcoholic liquid of lower temperature than the vapors themselves, and in very small quantity, the vapor of water will be partly condensed, so that the remainder will be richer in alcohol than it was previously. But the water, in condensing, converts into vapor a portion of the spirit contained in the liquid interposed, so that the uncondensed vapors passing away are still further enriched by this means. Here, then, are the results obtained; the alcoholic vapors are strengthened, firstly, by the removal of a portion of the water wherewith they were mixed; and then by the admixture with them of the vaporized spirit placed in the condenser. By the employment of some such method as this, a very satisfactory yield of spirit may be obtained, both with regard to quality, as it is extremely concentrated, and to the cost of production, since the simple condensation of the water is made use of to convert the spirit into vapor without the necessity of having recourse to fuel. The construction of every variety of distilling apparatus now in use is based upon the above principles.

A sectional view of another simple form of still is shown in Fig. 8; *V* is a wooden vat having a tight fitting cover *a*, through the center of which a hole has been cut. The wide end of a goose neck of copper pipe *g* is securely fitted over this aperture, the smaller end of this pipe passes through the cover of the retort *R* extending nearly to the bottom; *f* is the steam supply pipe from boiler; *M* the rectifier consisting of a cylindrical copper vessel containing a number of small vertical pipes surrounded by a cold water jacket; *o* the inlet for the cold water which circulates around these small pipes, discharging at *n*; the pipes in *M* have a common connection to a pipe *p*, which connects the rectifier with coil in cooler *C*; *s* is a pipe to the receptacle for receiving the distillate; *u* cold

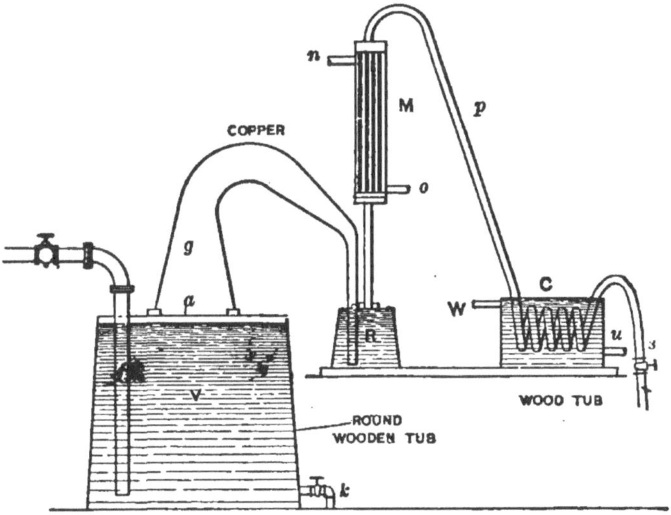

FIG. 8.—Simple Still, with Rectifier.

water supply pipe to cooler, and *W* discharge for warmed-up water, *k* discharge for refuse wash in vat *V*.

The operation is as follows: The vat *V* is nearly filled with fermented mash and retort *R* with weak distillate from a previous operation. Steam is then turned into the pipe *f* discharging near the bottom of the vat *V* and working up through the mash. This heats up the mash and the vapors escape up *g* over into *R* where they warm up the weak distillate. The vapors thus enriched rise into *M*, where a good percentage of the water vapor is distilled, that is, condensed by the cold water surrounding the small pipes. The vapor then passes over through *p* into the coil, where it is liquified and from whence it passes by pipe *s* into the receiver. The cold water for cooling both *M* and *C* can be turned on as soon as the apparatus has become thoroughly heated up.

The stills in use to-day in many parts of the South for the production of whiskey are quite as simple as those above described, and some for the making of " moonshine " liquor are more so.

The first distilling apparatus for the production of strong alcohol on an industrial scale was invented by Edward Adam, in the year 1801. The arrangement is shown in Fig. 9, in which *A* is a still to contain the liquor placed over a suitable heater. The vapors were conducted by a tube into the egg-shaped vessel *B*, the tube reaching nearly to the bottom; they then passed out by another tube into a second egg *C*; then, in some cases, into a third, not shown in the figure, and finally into the worm *D*, and through a cock at *G* into the receiver. The liquor condensed in the first egg is stronger than that in the still, while that found in the second and third is stronger than either. The spirit which is condensed at the bottom of the worm is of a very high degree of strength. At the bottom of each of the eggs, there is a tube connected with the still, by which the concentrated liquors may be run back into *A* for redistillation after the refuse liquor from the first distill has been run off.

In the tube is a stop-cock *a*, by regulating which, enough liquor could be kept in the eggs to cover the lower ends of the entrance pipes, so

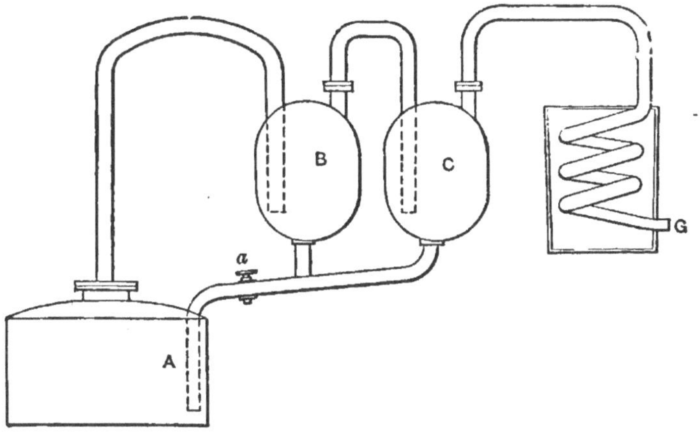

FIG. 9.—Adam's Still.

that the alcoholic vapors were not only deprived of water by the cooling which they underwent in passing through the eggs, but were also mixed with fresh spirit obtained from the vaporization of the liquid remaining in the bottom of the eggs, in the manner already described.

Adam's arrangement fulfilled, therefore, the two conditions necessary for the production of strong spirit inexpensively; but unfortunately it had also serious defects. The temperature of the egg could not be maintained at a constant standard, and the bubbling of the vapors through the liquor inside created too high a pressure. It was, however, a source of great profit to its inventor for a long period, although it gave rise to many imitations and improvements.

The operation of distilling is often carried on in the apparatus represented in **Fig. 10.** It is termed the Patent Simplified Distilling Apparatus; it was originally invented by Corty, but it has since undergone much improvement. *A* is the body of the still, into which the wash is put; *B* the head of the still; *c c c* three copper plates fitted in the upper part of the three boxes; these are kept cool by a supply of water from the pipe *E*, which is distributed on the top of the boxes by means of the pipes G G G. The least pure portion of the ascending vapors is condensed as it reaches the lowest plate, and falls back, and the next portion as it reaches the second plate, while the purest and lightest vapors pass over the goose-neck, and are condensed in the worm, The temperature of the plates is regulated by altering the flow of water by means of the cock *F*. For the purpose of cleaning the apparatus, a jet of steam or water may be introduced at *a*. A regulator is affixed at the screw-joint *H*, at the lower end of the worm, which addition is considered an important part of the improvement. The part of the apparatus marked *I* becomes filled soon after the operation has commenced; the end of the other pipe *K* is immersed in water in the vessel *L*. The advantage claimed for this apparatus is that the condensation proceeds in a partial vacuum, and that there is therefore a great saving in fuel. One of these stills, having a capacity of 400 gallons, is said to work off four or five charges during a day of 12 hours, furnishing a spirit 35 per cent. over-proof.

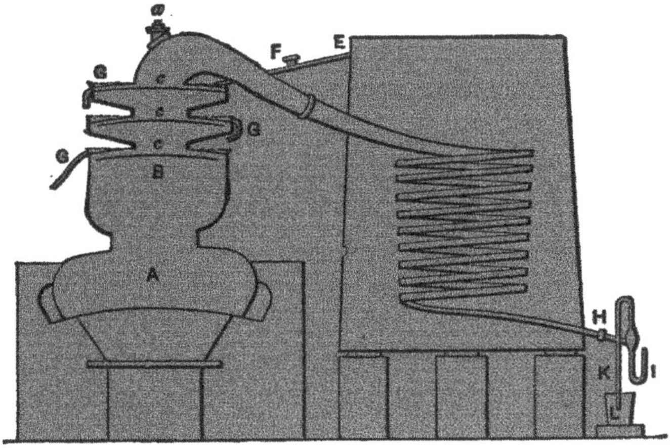

FIG. 10.—Corty's Simplified Distilling Apparatus.

Fig. 11 represents a double still which was at one time largely employed in the colonies. It is simply an addition of the common still *A* to the patent still *B*. From time to time the contents of *B* are run off into *A*, those of *A* being drawn off as dunder, the spirit from *A* passing over into *B*. Both stills are heated by the same fire; and it is said that much fine spirit can be obtained by their use at the expense of a very inconsiderable amount of fuel.

Compound Distillation. Where stills of the form shown in Figs. 6 and 8 are used the alcohol obtained is weak. Hence it is necessary that the distillate be again itself distilled, the operation being

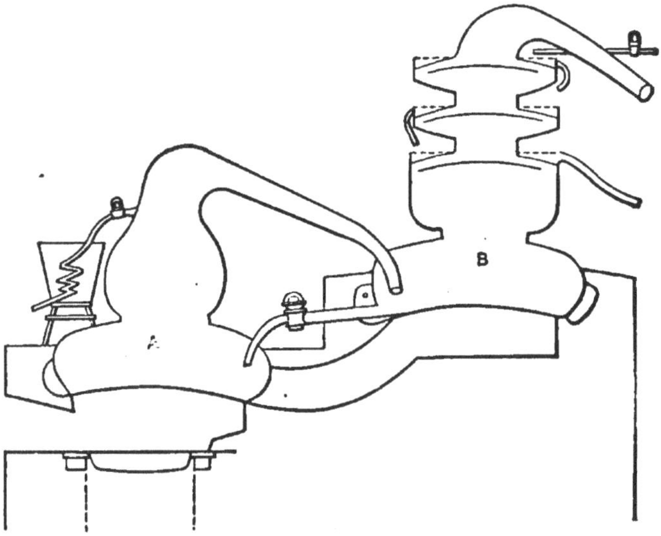

FIG. 11.—Double Still.

repeated a number of times. In the better class of still, however, compound distillation is performed the mash is heated by the hot vapors rising from the still and the vapors are condensed and run back into the still greatly enriched.

The principle of compound distillation is well shown in Dorn's apparatus, Fig. 12. This consists

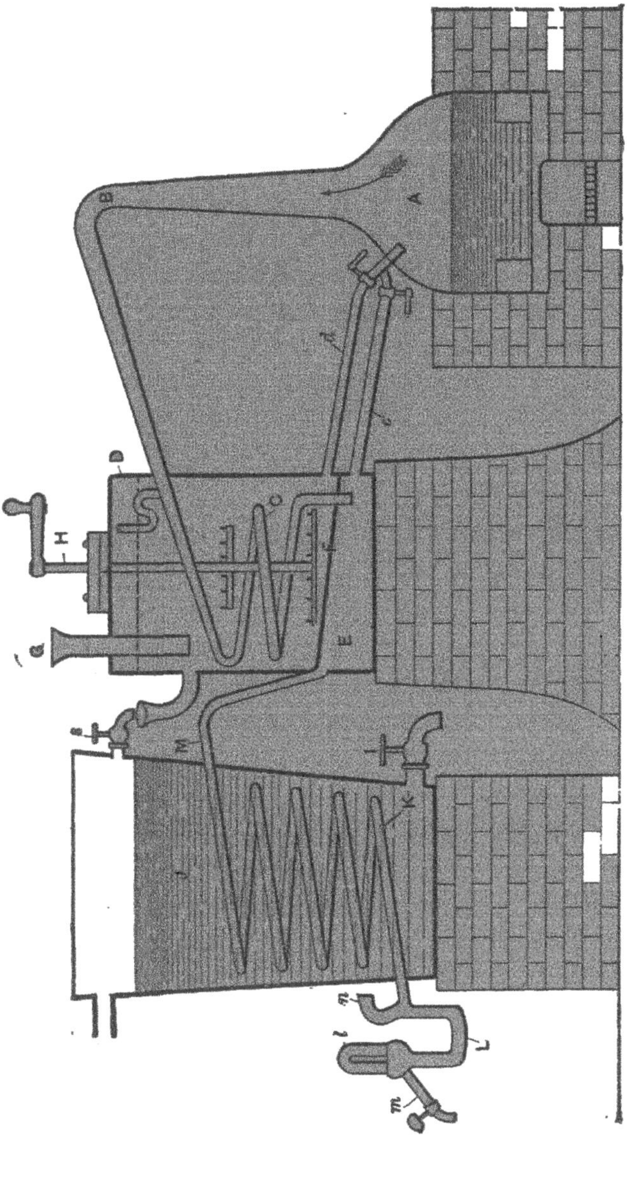

FIG. 12.—Dorn's Compound Still.

of a still or boiler *A* having a large dome-shaped head, on the interior faces of which the alcoholic vapors will condense. Thus only enriched vapors will pass up through goose-neck *B* to the mash heater *D*. *C* is a worm the end of which passes out to a compartment *E* through an inclined partition *F*. From the compartment *E* a pipe *e* leads into the still *A*. An agitator *H* is used for stirring the mash, so that it may be uniformly heated. A pipe *d* provided with a cock allows the mash to be drawn off into the still *A*. From the highest point of the compartment *E* a pipe *M* leads to condensing coil *K* in a tub *J* of cold water, having a draw-off cock *I*.

At the exit end of the condensing worm *K* the tube is bent in a U form as at *L*, one arm of which has a curved open-ended continuation *n*, through which the air in the worm is expelled. The other arm opens into an inverted jar *l* containing a hydrometer, for indicating the strength of the spirit. The spirits pass off through *m* into a receiver.

In operation the mash is admitted into the heater *D* through *G* until the heating tank is nearly filled. A certain amount of mash is then allowed to run into the still *A* through the pipe *d*. The cock in *d* is closed and the fire lighted.

The vapors from the still are condensed in worm *C* and the condensed liquid drops down into compartment *E*. Any vapor passing through *B* and *C* so highly heated as to be uncondensed in coils *C* passes through the layer of liquid in compartment *E*, collects in the highest portion of the compartment and passes through pipe *M* to coil *K* where it is entirely liquefied. If the liquid in *E* rises beyond a certain level it passes through pipe *e* back to the still. Any vapors which may collect in the upper part of *D* pass into the small bent pipe opening into the first coil of worm *C*. Water for rinsing the heater *D* may be drawn through cock *s* from the tub *J* and warm water for rinsing the still, through pipe *d* from the heater.

Another form of compound still is shown in Fig. 13. In this the still *S* is divided into an upper and lower compartment by a concavo-convex partition *d*, having at its crown an upwardly extending tube *t* from which projects side tubes *p*. A pipe *P* opens above and extends from tube *t*. *C* is the mash heater and condenser. Connected to the head of the still is a pipe *T* through which the vapors pass to a condensing coil *f* formed on the wall of the heater *C*. At its bottom the coil *f* extends out of the heater, through the water tub *W* and out to receiver as at *F*. In the head of this heater is a valve *V* whereby any vapors which may arise from the heated mash are conducted by pipe *U* to *T*.

The heater *C* is filled through funnel *Y* and the mash is admitted to the still through pipe *b* having cock *a*. The pipe *P* extends to the upper part of the water tub *W* and then downward to the bottom, where it again enters the still.

An opening in the partition *d* is controlled by a valve *G* which allows liquid in the upper compartment of the still to flow into the lower. Spent mash may be drawn off through *c* and the height of the water in tub *W* be regulated by pipe *Z*.

FIG. 13.—Compound Still.

The operation of this still is similar to Dorn's still. Mash is put into *C* and a quantity of it is let into the upper compartment of the still and into the lower compartment by valve *G*. This valve is closed and the fire started. The vapors pass upward through *t*. If they are quite highly vaporized they pass onward up *P*, are condensed in their passage through the cool water tub and return as liquid to the upper compartment where they are further heated.

The liquid in the upper compartment is thus constantly enriched and the vapor therefrom passes out through pipe *T* into condensing coils

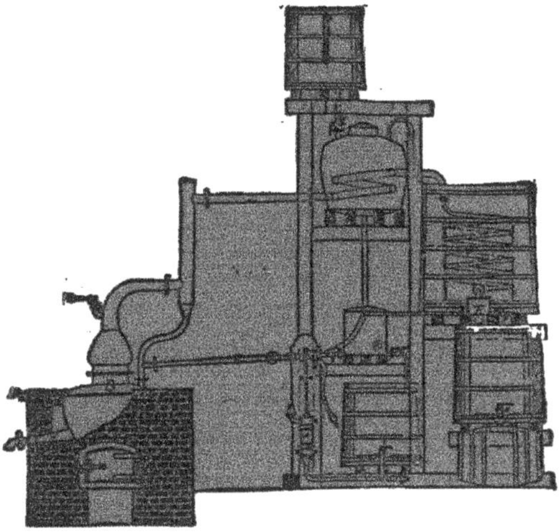

FIG. 14.—Compound Direct Fire Still.

f where it is condensed into spirit and passes off by *F*.

The funnel tube *Y* acts also as a means of warning the attendant as to the condition of the mash. If it is too high in level and the pressure of vapor in the heater *C* too great, liquid will be forced out of *Y*; if on the contrary, the mash sinks below the level of the pipe then vapor will escape and the heater needs refilling.

Fig. 14 shows a simple form of compound direct fire still as manufactured by the Geo. L. Squier Mfg. Co., of Buffalo, N. Y.

Cellier-Blumenthal carrying this principle further devised an apparatus which has become the basis of all subsequent improvements; indeed, every successive invention has differed from this arrangement merely in detail, the general principles being in every case the same. The chief defect in the simple stills was that they were intermittent that is required the operations to be suspended when they were recharged, while that of Cellier-Blumenthal is continuous; that is to say, the liquid for distillation is introduced at one end of the arrangement, and the alcoholic products are received continuously, and of a constant degree of concentration, at the other. The saving of time and fuel resulting from the use of his still is enormous. In the case of the simple stills, the fuel consumed amounted to a weight nearly three times that of the spirit yielded by it; whereas, the Cellier-Blumenthal apparatus reduces the amount to one-quarter of the weight of alcohol produced. Fig. 15 shows the whole arrangement, and Figs. 16 to 17 represent different parts of it in detail.

In Fig. 15 *A* is a boiler, placed over a brick furnace; *B* is the still, placed beside it, on a slightly higher level and heated by the furnace flue

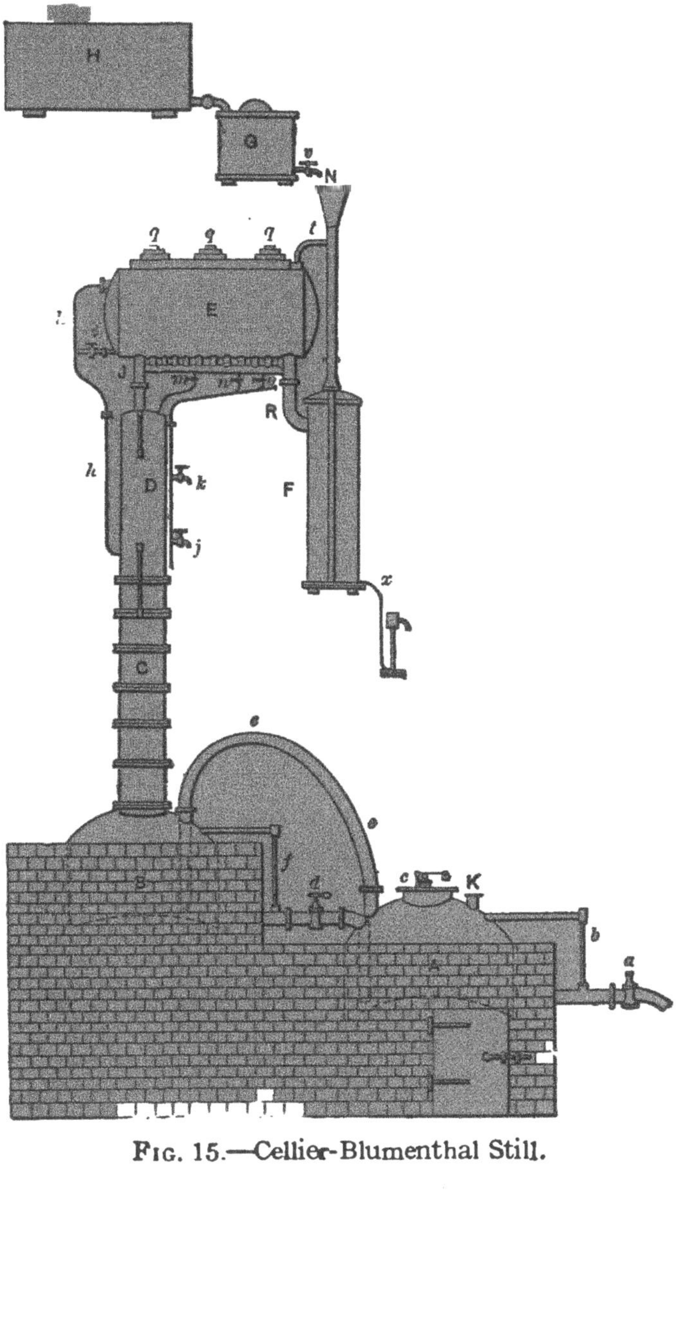

Fig. 15.—Cellier-Blumenthal Still.

which passes underneath it. A pipe *e* conducts the steam from the boiler to the bottom of the still. By another pipe *d*, which is furnished with a stop cock and which reaches to the bottom of the still *A*, the alcoholic liquors in the still may be run from it into the boiler; by turning the valve the

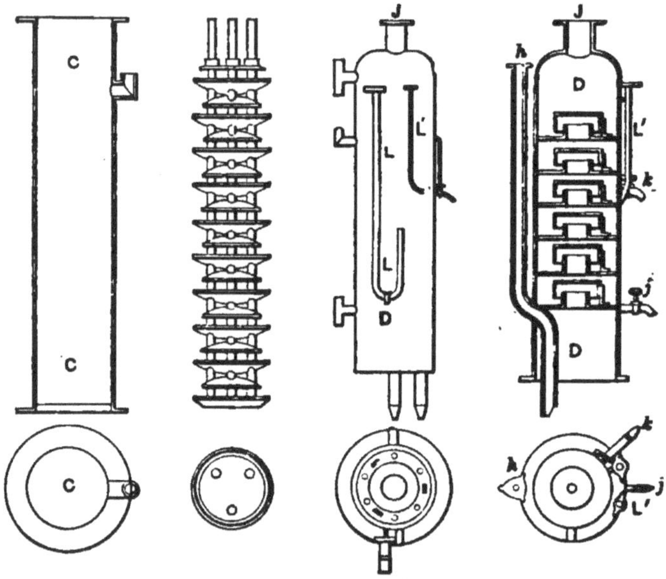

FIG. 16.—Details of Rectifier Column.

spent liquor may be run out at *a*. The glass tubes *b* and *f* show the height of liquid in the two vessels. *K* is the valve for filling the boiler and *c* the safety valve.

The still is surmounted by a column *C*, shown in section in Fig. 16. This column contains an enriching arrangement whereby the liquid flowing down into the still *B* is brought into intimate contact with the steam rising from the still. The liquid meets with obstacles in falling and falls downward in a shower, which thus presents multiplied obstacles to the ascent of the vapor. The liquid is thus heated almost to the boiling point before it falls into the still *B*. The construction for effecting this is shown at *C*, Fig. 16 and consists of an enclosed series of nine sets of circular copper saucer-shaped capsules, placed one above the other, and secured to three metallic rods passing through the series so that they can be all removed as one piece. These capsules are of different diameters, the larger ones which are, nearly the diameter of the column, are placed with the rounded side downwards, and are pierced with small holes; the smaller ones are turned bottom upwards, a stream of the liquid to be distilled flows down the pipe *h* from *E*, into the top capsule of *C* and then percolating through the small holes, falls into the smaller capsule beneath, and from the rim of this upon the one next below, and so throughout the whole of the series until it reaches the bottom and falls into the still *B*. The vapors rise up into the column from the still and meeting the stream of liquid convert it partially into vapor which passes out at the top of *C* considerably enriched, into the column *D*.

Fig. 16 shows a sectional view of the column *D*, the "rectifying column" as it is called. It contains six vessels, placed one above the other.

in an inverted position, so as to form seals. These are so disposed that the vapors must pass through a thin layer of liquor in each vessel. Some of the vapor is thus condensed and the condensed liquid flows back into column *C*, the uncondensed vapor considerably enriched passing up the pipe *J*, into the coil *S* in the condenser *E*, Fig. 17, which is filled with the "wash" to be distilled.

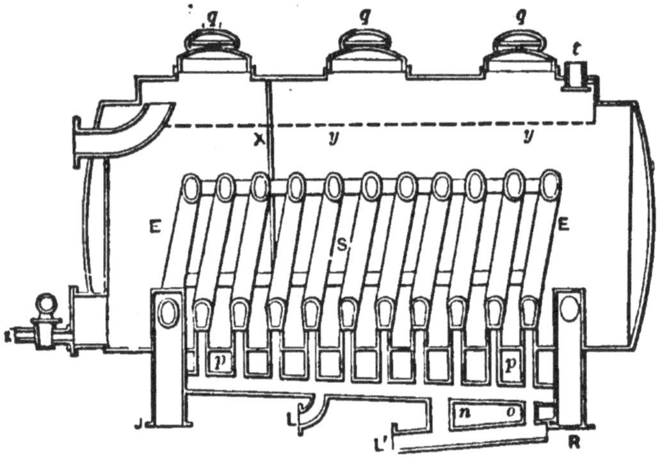

FIG. 17.—Details of Condenser and Mash Heater.

Entering by the pipe *t*, Fig. 15, the undistilled liquid or "wash" is distributed over a perforated plate *y y*, and falls in drops into the condenser *E*, where it is heated by contact with the coil *S* containing the heated vapors. The condenser is divided into two compartments by a diaphragm *X* which is pierced with holes at its lower extremity;

through these holes the wash flows into the second compartment, and passes out at the top, where it runs through the pipe *h*, into the top of the column *C*.

The vapors are made to traverse the coil *S*, which is kept at an average temperature of 122° F., in the right hand compartment, and somewhat higher in the other. They pass first through *J* into the hottest part of the coil, and there give up much of the water with which they are mixed, and the process of concentration continues as they pass through the coil. Each spiral is connected at the bottom with a vertical pipe by which the condensed liquors are run off; these are conducted into the retrograding pipe *p p*. Those which are condensed in the hottest part of the coil, and are consequently the weakest, are led by the pipe *L* into the third vessel in the column *D*, Fig. 16, while the stronger or more vaporized portions pass through *L′* into the fifth vessel. Stop-cocks at *m*, *n*, *o* regulate the flow of the liquid into these vessels, and consequently also the strength of the spirit obtained.

Lastly, as the highly concentrated vapors leave the coil *S* at *R*, they are condensed in the vessel *F*, which contains another coil. This is kept cool by a stream of liquid flowing from the reservoir *H* into the smaller cistern *G* from which a continuous and regular flow is kept up through the tap *v* into a funnel *N* and thence into condenser *F*. It ultimately flows into condenser *E* through pipe *t*, there being no other outlet. The finished products run out by pipe *x* into suitable receivers.

It will be seen that the condenser *E* has two functions. First it condenses the alcoholic vapors before transmitting them to the final condenser *F*, rejecting and sending back those vapors which are not highly enough vaporized. Second it heats the wash intended for distilling by appropriating the heat of the vapors to be condensed. Thus two birds are killed with one stone. It will be noticed that the same result is accomplished in the columns *C* and *D*. This is the principle of all modern stills.

Another form of still which is very analogous to that last described is Coffey's apparatus, shown in Fig. 18, and is the immediate prototype of the stills used to-day in all but the simplest plants.

It consists of two columns, *C* the analyser, and *H* the rectifier, placed side by side and above a chamber containing a steam pipe *b* from a boiler *A*. This chamber is divided into two compartments by a horizontal partition *a* pierced with small holes and furnished with four safety valves *e e e e*. The column *C* is divided into twelve small compartments, by means of horizontal partitions of copper, also pierced with holes and each provided with two little valves *f*. The spirituous vapors passing up this column are led by a pipe *i* to the bottom of the second column or *rectifier*. This column is also divided into compartments in precisely the same way, except that there are

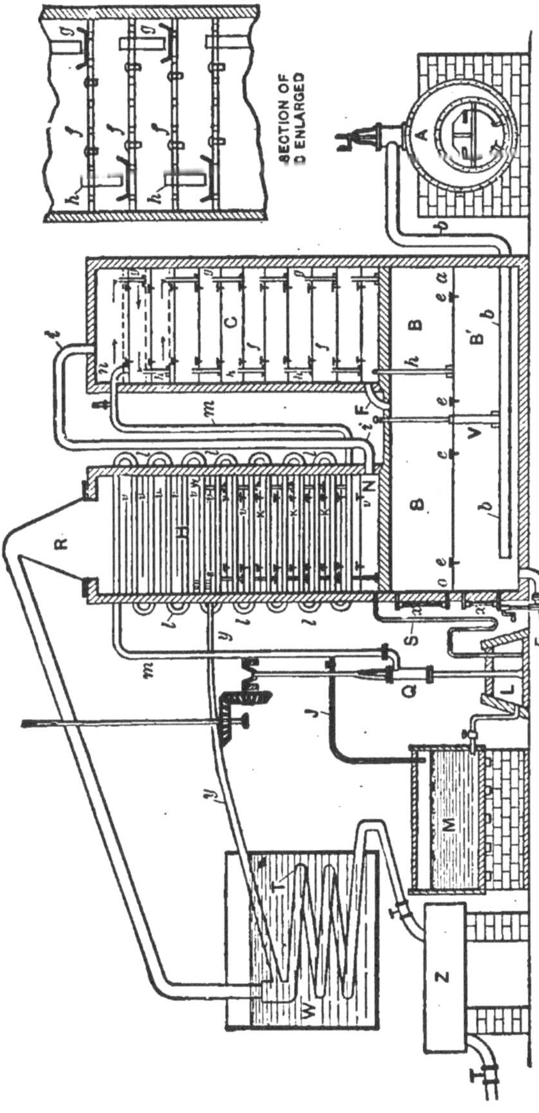

FIG. 18.—Coffey's Rectifying Still.

fifteen of them, the ten lowest being separated by the partitions, which are pierced with holes. The remaining five partitions are not perforated, but have a wide opening as at *w*, for the passage of the vapors, and form a condenser for the finished spirit. Between each of these partitions passes one bend of a long zig-zag pipe *m*, beginning at the top of the column, winding downwards to the bottom, and finally passing upwards again to the top of the other column, so as to discharge its contents into the highest compartment. The apparatus works in the following way: The pump *Q* is set in motion, and the zig-zag pipe *m* then fills with the wash or fermented liquor until it runs over at *n* into the highest compartment of column *C*. The pump is then stopped, and steam is introduced through *b*, passing up through the two bottom chambers and the short pipe *F* into the analyzing column, finally reaching the bottom of the other column by means of the pipe *i*. Here it surrounds the coil pipe *m* containing the wash, so that the latter becomes rapidly heated.

When several bends of the pipe have become heated, the pump is again set to work, and the hot wash is driven rapidly through the coil and into the analyzer at *n*. Here it takes the course indicated by the arrows, running down from chamber to chamber through the tubes *h* until it reaches the bottom; none of the liquor finds its way through the perforations in the various partitions, owing to the pressure of the ascending steam.

As the liquid cannot pass through the holes in the partitions it can only pass downward through the drop-pipe tubes *h*. By this means the mash is spread in a thin stratum over each partition to the depth of the seal *g* and is fully exposed to the steam forcing its way up through the holes, the alcohol it contains being thus volatilized at every step.

In its course downwards the wash is met by the steam passing up through the perforations, and the whole of the spirit which it contains is thus converted into vapor. As soon as the chamber B is nearly full of the spent wash, its contents are run off into the lower compartment by opening a valve in the pipe *V*. By means of the cock *E*, they are finally discharged from the apparatus. This process is continued until all the wash has been pumped through.

The course taken by the steam will be readily understood by a glance at the figure. When it has passed through each of the chambers of the analyzer, the mixed vapors of water and spirit pass through the pipe *i* into the rectifying column. Ascending again, they heat the coiled pipe *m*, and are partially deprived of aqueous vapors by condensation. Being thus gradually concentrated, by the time they reach the opening at *w* they consist of nearly pure spirit, and are then condensed by the cool liquid in the pipe, fall upon the partition and are carried away by the pipe *y* to a refrigerator *W*. Any uncondensed gases pass out by the pipe *R* to the same refrigerator, where they are deprived of any alcohol they may contain. The weak liquor condensed in the different compartments of the rectifier descends in the same manner as the wash descends in the other column; as it always contains a little spirit, it is conveyed by means of the pipe *S* to the vessel *L* in order to be pumped once more through the apparatus.

The condensed spirit gathered over the plates *v* passes out through the pipe *y* to the condensing worm *T*. If any vapors escape the condensing plates they pass into *R* and are condensed in the worm *T* also. From worm *T* the spirit flows into a suitable receiver *Z*.

Before the process of distillation commences, it is usual, especially when the common Scotch stills are employed, to add about one lb. of soap to the contents of the still for every 100 gallons of wash. This is done in order to prevent the liquid from boiling over, which object is effected in the following way: The fermented wash always contains small quantities of acetic acid; this acts upon the soap, liberating an oily compound which floats upon the surface. The bubbles of gas as they rise from the body of the liquid are broken by this layer of oil, and hence the violence of the ebullition is considerably checked. Butter is sometimes employed for the same purpose.

Figs. 19 and 20 show a diagrammatic section and a plan of a still used for thick mashes which are liable to burn. This comprises a circular chamber *B* supported over suitable heating means, having on its bottom a series of concentric partitions *b* which divide the bottom of the chamber into shallow channels for the mash. Running diametrically through the chamber is a partition.

The mash passes from a tank as *A* by a passage *a* to an opening on one side of the central portion and into the outside channel *b*. The current of

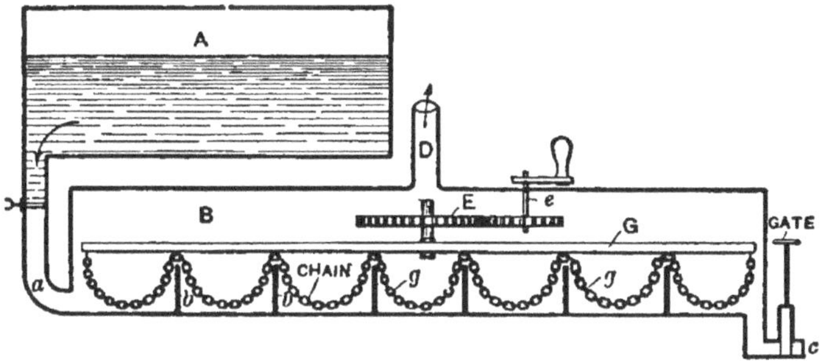

FIG. 19.—Rotary Current Still.

liquid passes along the outer channel until it is deflected by the central partition into the next interior channel *b* and so on until it arrives at the center when it passes through the central partition into the other half of the chamber. Here it passes around back and forth and gradually outward to the outermost channel from which it passes off through an adjustable gate in outlet *c*. By adjusting this gate, and a gate or cock in inlet passage *a*, the passage and consequent depth of the liquid in the channels may be regulated. The vapor rising from the mash is carried over to a condenser through pipe *D*. In order to keep the mash from burning a chain *g* is rotarily reciprocated along the channels by means of the bar *G*, the gear *E* and the crank shaft *e*. Various modifications of this construction have been devised. The advantage of the still lies in submitting the mash in a thin current to the action

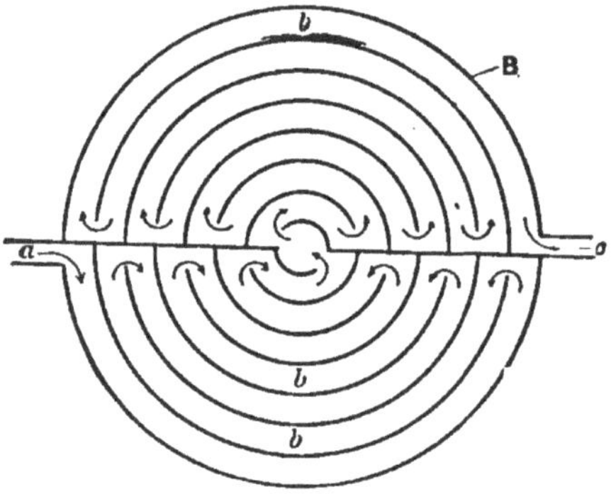

FIG. 20.—Rotary Current Still.

of the heat, and the consequent rapid vaporization.

Every distillation consists of two operations: The conversion of liquid into vapor, and the reconversion of the vapor into liquid. Hence perfect equilibrium should be established between the vaporizing heat and the condensing cold. The quantity of vapor must not be greater or less than can be condensed. If fire is too violent the vapors will pass out of the worm uncondensed. If the fire is too low the pressure of the vapor is not great enough to prevent the entrance of air, which obstructs distillation. As a means of indicating the proper regulation of the fire, the simple little device shown in Fig. 21 may be used.

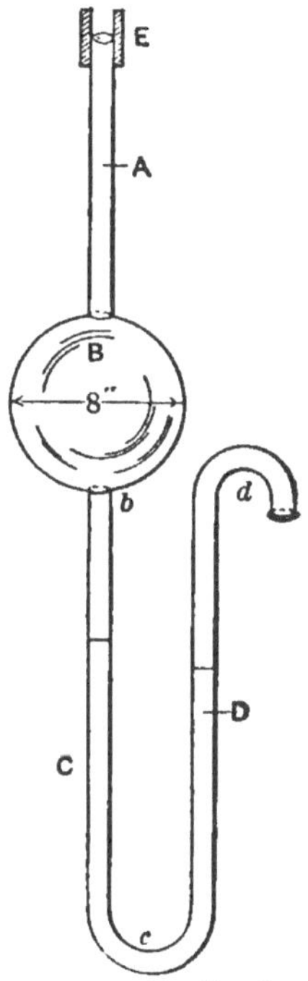

FIG. 21.—Indicator for Regulating the Distilling Fire.

This consists of a tube of copper or glass having a ball *B* eight inches in diameter. The upper end *E* of the tube is attached to the condensing worm. The lower end of the tube is bent in U-shape; the length of the two bends from *b* to outlet is four feet. The ball has a capacity slightly greater than the two legs of the bend.

Normally the liquid in the two legs will stand at a level. If, however, the fire is too brisk the vapor will enter the tube and drive out the liquor at *d*, and thus the level in the leg *C* will be less than in the leg *D*. If, however, the fire is low, the pressure of vapor in the worm will decrease and the pressure of the outside air will force down the liquid in leg *D* and up leg *C* into the ball.

A more perfected device but operating on the same principle is shown in Fig. 26.

It is obviously impossible to present in the small compass of this book a description of all the varieties of stills used, but these which have been described illustrate the principles on which all stills are constructed and were chosen for their simplicity of construction and clearness of their operation. The principle of their operation is exactly the same as the more modern forms now to be described.

CHAPTER IV.

MODERN DISTILLING APPARATUS.

In the previous chapter we have given a description of small, simple stills, such as were used until late years, and which are yet used in many localities where distilling is carried on on a small scale. We will now describe the principle features of more complicated and elaborate apparatus.

All modern distilling apparatus for the production of a high grade of alcohol is based upon the principle set forth in the description of the Coffey still; that is, upon using a distilling column and a concentrating column, wherein the "wash" or mash fermented as described, passes over a series of plates or other obstructions in contact with an ascending column of heated vapor. This heated vapor extracts the alcohol from the wash, or from the low wines of the concentrator, and is continually strengthened during its journey until it passes off to a condenser as a vapor very rich in alcohol. The converse of this is true with the wash, which in its downward course is gradually deprived of its alcohol until it finally passes off at the bottom of the column.

Fig. 22 is illustrative of the general form and

arrangement of such a column and its adjuncts; the details, however, will vary with each make of still. In this the "column" consists of a casing really continuous but divided into two portions—the distilling portion *A* and the rectifying portion

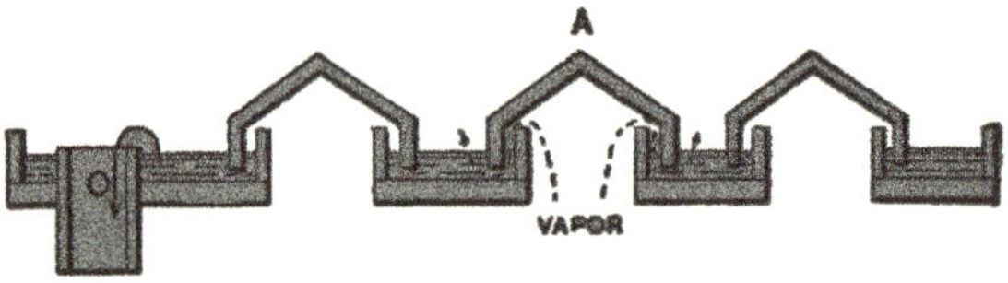

Cross-Section of Fig. 23.

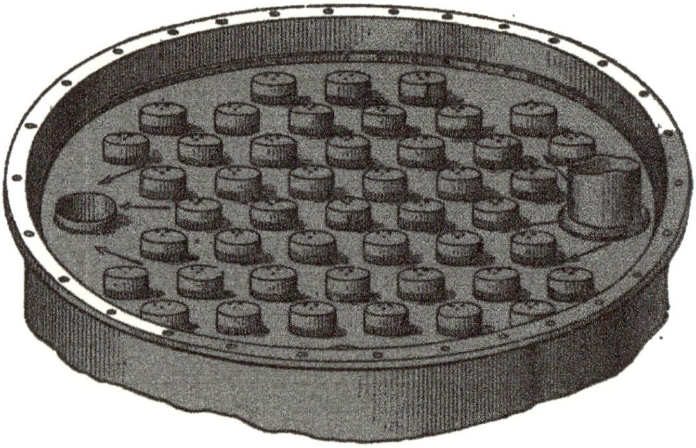

Fig. 23.—Distilling Plate.

B. The operation is alike, however, in principle in both portions.

The wash by means of a suitable pump is forced into an overhead tank or concentrator *G* where it is warmed by the hot vapors as will be later de-

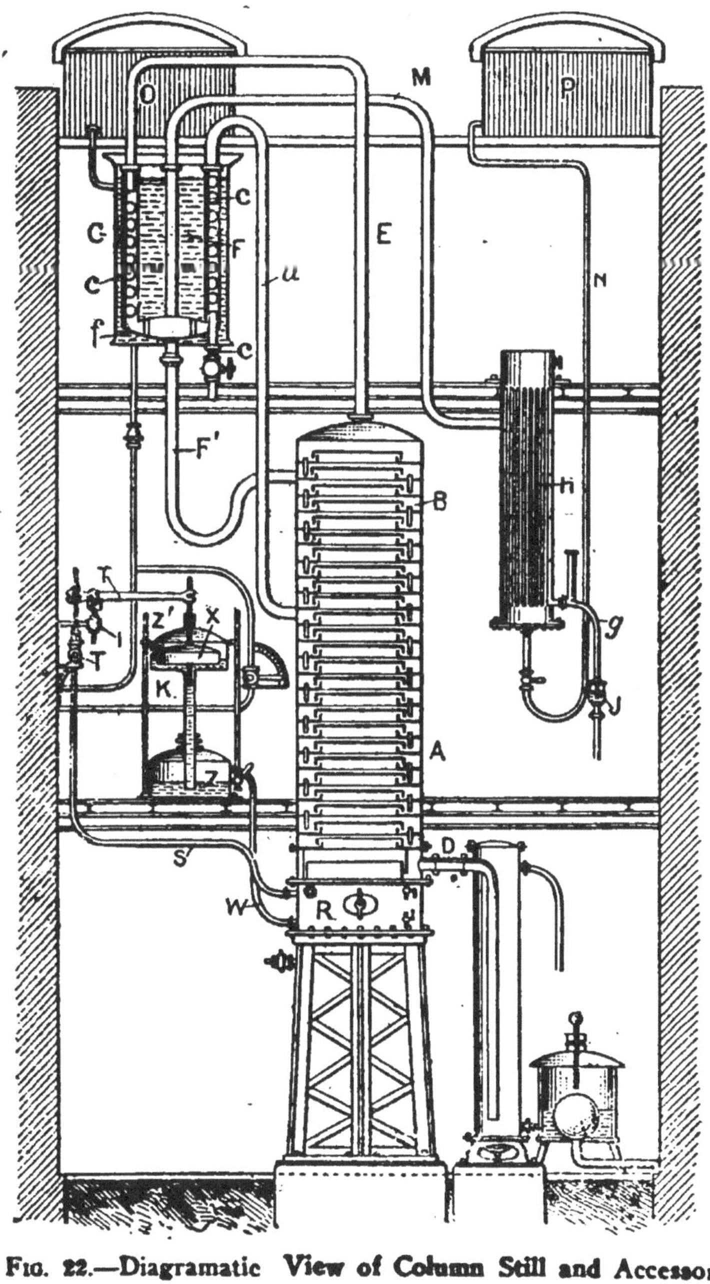

Fig. 22.—Diagramatic View of Column Still and Accessory Apparatus.

(To face page 64)

scribed. It passes around the interior of the concentrator in a coil *c* and then passes off by a pipe *c* to the uppermost plate of the distilling portion *A* of the column.

The plates, as before explained on page 55, are each formed with a dropping tube *O* (see Fig. 23), which extends above the plate to an extent slightly less than the desired thickness of the layer of liquid on each plate, and with perforations each having an upwardly projecting rim, and each covered with a cap *A*. This rim and cap form a trap. The ascending vapors pass up through the perforations, down between the rim and the edge of the cap and thus out through the layer of wash contained on the cap. The wash remains constantly level with the top of the tube *O*, the excess running off through the tube *O* to the compartment or plate beneath.

To return to Fig. 22, the wash by the pipe *a* enters the distilling portion of the column at the uppermost plate thereof and, as described above, drops down from plate to plate. A steam pipe *S* enters the bottom compartment of the distilling portion of the column and the steam as it rises through the little traps, bubbles out through the layer of wash and in each compartment enriches itself with alcohol. Thus the rising column of vapor is constantly becoming richer and the downward current of wash constantly weaker until at last it passes away as spent wash at the very bottom of the column by the pipe *D*.

The hot vapors, as before described, pass upward and enter the rectifying portion of the column *B*. This consists of a series of compartments having perforated bottoms and dropping tubes. The vapor passes upward through these perforations of the plates,—the condensed portion of it dropping back again on to the lower plates or on to the distilling plates to be again vaporized and concentrated and the more highly vaporized portion passing out at the top of the column through the pipe *E* to the concentrator *G*.

The concentrator consists of a tank containing water within which is supported a vessel *F* having double walls. The interior of this vessel is likewise filled with water. Between the double walls and surrounding the coiled pipe *c* passes the vapors from pipe *E*.

At the bottom of the vessel *F* is a compartment *f* connected by a pipe F^1 with the upper compartment of the rectifying column. The less highly heated vapors will be condensed by the passage through the double walls of the vessel and the condensation will collect in the compartment *f*, and from there pass off by pipe F^1 back to the rectifying column, to be again vaporized and strengthened by the descent from plate to plate of *B*.

The rich and highly vaporized vapors which have passed the test of this preliminary concentration, pass out of the compartment *f* by a pipe *M*. Here again the water surrounding the pipe tends to condense all but the most highly charged vapor and send it back to compartment *f* but the vapor which succeeds in passing over through pipe *G* is carried downward to a condenser *H* where it is finally condensed and drawn off as at *g*. It is necessary that the rate of mash feed be regulated so that neither too much mash shall be pumped into the mash heater *G*, or too little, and the pipe leading from the pump to the heater is therefore provided with a tap and an indicating dial.

In these modern stills the following are particularly important points to be especially brought to the consideration of the distiller.

It cannot be too strongly impressed that effectiveness of the distilling column depends on the plates dividing it,—that is, upon the horizontality of the plates and the form of the traps or perforations. If the plates are not horizontal the wash is not maintained at a uniform level across the entire extent of the plate and hence some of the ascending vapor will pass out without contacting with the wash through uncovered traps, while others of the traps will be so deeply submerged in wash that the vapor cannot bubble through.

Again the caps should be so made as to divide the vapor into fine streams and bring it into contact with each part of the wash. Plates simply perforated and uncapped give excellent results for they molecularize the vapor ascending through the liquid contained on the plates, but they require a constant pressure of vapor, and any variations of pressure tends to discharge them. In addition these perforations gradually enlarge by the action of acids in the wash or clog up, and the apparatus soon works badly.

Good forms of capped traps are those shown in Figs. 24, 25 devised by Barbet. These are provided with an interior upwardly projecting rim. Extending over the rim and down around it is a copper cap having its margin slitted.

The wash carried on the plate circulates about the caps and the alcoholic vapors bubble out

Fig. 24.—Barbet Trap.

Fig. 25.—Barbet Trap.

through the slits and up through the wash, the vapor thus being finely divided and coming into intimate contact with each portion of the wash and thus more thoroughly depriving it of its aclohol.

Besides this there is another advantage resident in these caps, namely, that distillation may be stopped for several hours and then re-started without trouble for the reason that the wash has been retained on the plates, whereas were the plates simply perforated the wash would ooze through and the plates have to be recharged. This form of plate may be easily repaired and does not necessitate the removal or replacement of the plate itself. The caps alone need be removed.

For thick washes, which tends to obstruct the slits of the cap, Barbet has devised the cap shown at the right in Fig. 25. This cap extends down to the plate itself, and has very narrow slits in its periphery. With such a cap as shown in Fig. 24, the bran, sediments, etc., would tend to settle upon the top of the cap, enter beneath it and through the slits. The cone-shape of the top of this cap prevents the deposit of dregs thereon and the very narrow slits oppose the entrance of bran or sediment.

While, for the sake of clearness, an old form of concentrator, *G*, has been shown, the concentrator, preheater for the wash, and condensers, to-day, are usually composed of bundles of tubes through which the vapors pass surrounded by water or the cool wash. These should be of bronze or copper and made without solder. The tubes should be capable of being taken out for cleaning or repairing.

In many distilling apparatuses the distilling column and the rectifying column are in two parts, one beside the other. This overcomes the objection of having a very high column and also prevents the low wines, *i.e.*, the weak alcoholic liquor after its first concentration, from passing into the wash as it would do with the continucus column.

In order that the amount of steam entering the column may be regulated, the column is usually provided with a steam regulator (Fig. 26), whose principle of operation may be easily under stood by referring to Fig. 22. It comprises an upper and a lower chamber *Z Z'* connected by a central tube *K* which projects down nearly to the bottom of the lower chamber. A pipe *W* communicates with the steam chamber *R* of the column and enters the chamber *Z* above the level of the water contained therein. In the upper chamber *Z'*, is a

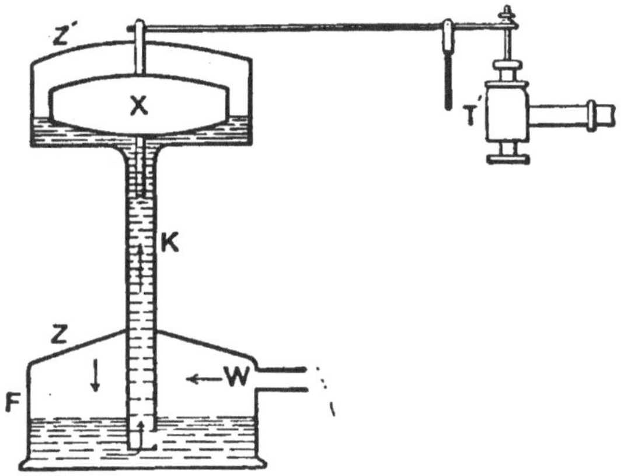

Fig. 26.—Steam Regulator.

float *X*, connected to the differential lever *T* of a steam valve *T'* which controls the inlet of steam passing through pipe *S* to the steam chest *R*. The principle of operation is very simple. When the pressure in the steam chest *R* becomes too great, steam in the pipe *W* and chamber *Z* forces the water therein up in tube *K*, thus lifting the float *X* and closing the steam entrance valve *T'*. When the pressure of steam is low, the level of the liquid in *Z* rises and liquid in *Z'* runs into *Z*, the float *X* falls. opening valve *T'* and allowing a greater flow of steam.

As it is often desirable to change the pressure of steam in the column at various points in the operation, the best regulators are usually provided with means to that end.

In order to measure the output of the still, there is attached thereto a gauge glass (*J* in Fig. 22), a diagram of which is shown in Fig. 27. This consists of a jar *A* connected at its lower end at *b* by an annular passage *B* to a chamber *E* from which proceed the taps *F*. Centrally through the passage *B* passes a tube *c* connected at its lower end to the pipe *C* leading from the condenser. The tube *C c* projects upward into the jar *A* and is open at its upper end.

Now the opening *b* is of a certain size and it is obvious that it will carry off a certain amount of liquid when running full or the amount allowed to flow out by the exit tap *F*. If now, more than that quantity of alcohol is produced, the alcohol will rise in the jar *A* until the rate of inflow and outflow is equal. If, however, the still is producing less than that quantity then the level of liquid in *A* will gradually drop. Hence, by observing the level of the liquid in *A* and its constancy or variation in level, it is possible to tell precisely how much alcohol is running per hour and if the rate is steady. The jar *A* is provided with a cap *G*

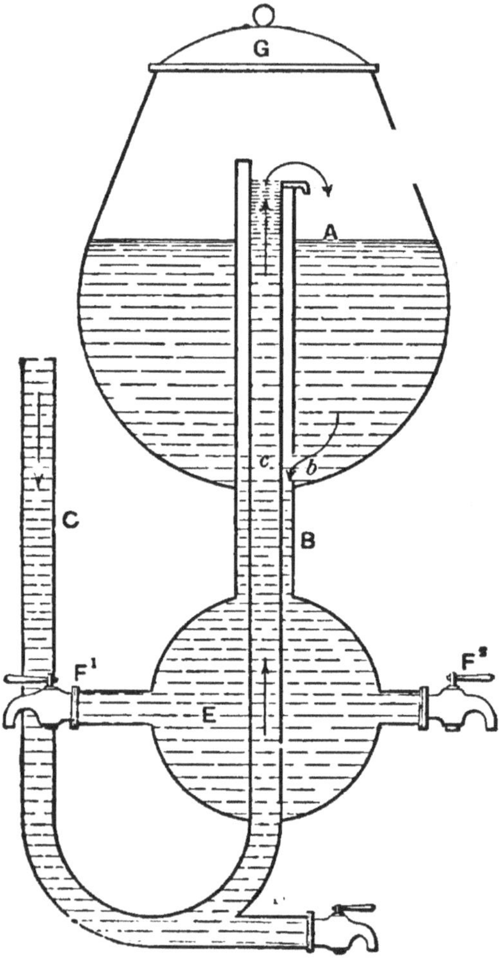

Fig. 27.—Gauge Glass for Regulatar.

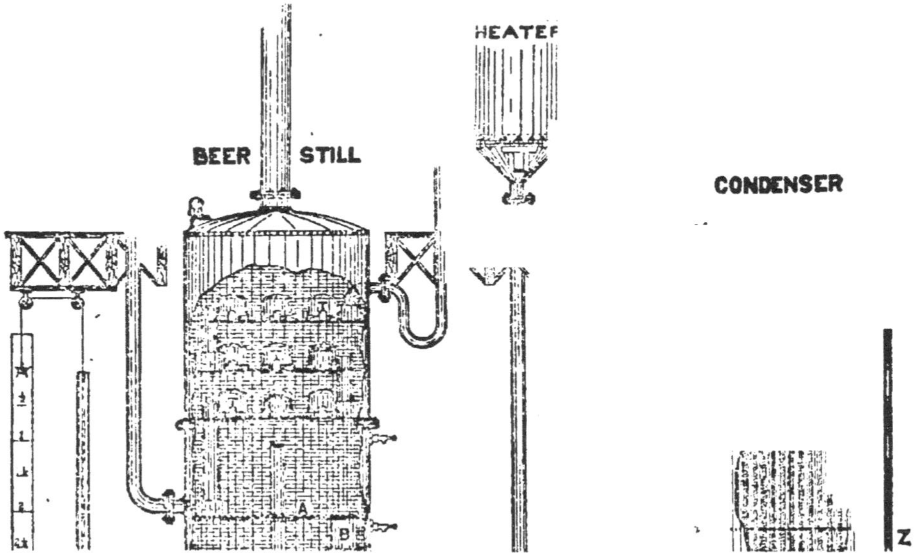
BEER STILL
HEATEF
CONDENSER
A
B
Z

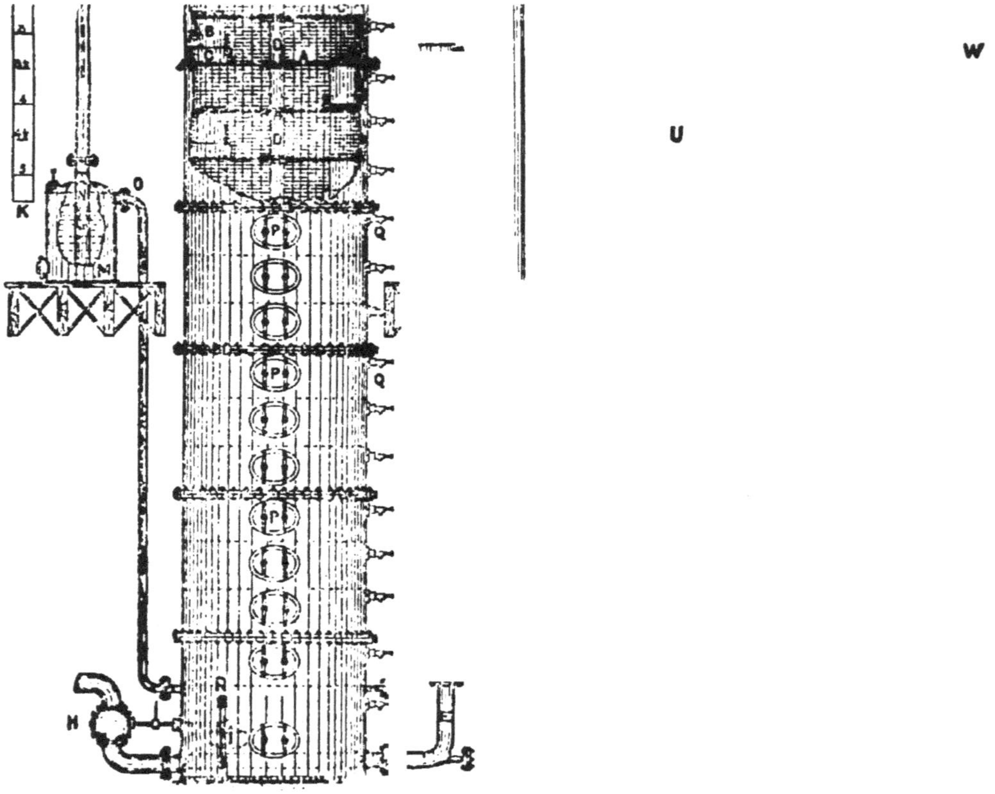

W
U
K
O
Q
Q
P
P
P
H
R

whereby an alcoholometer may be inserted into tube *c* for the purpose of testing the strength of the liquor. The taps *F* are for the purpose of collecting the first runnings, the pure alcohol and the last runnings or "feints."

These principles are also embodied in the ap-

Fig. 29—Detail of Chamber, Continuous Still.

paratus designed by the Vulcan Copper Works Co., of Cincinnati, and illustrated in Fig. 28. The apparatus comprises the still, a wash heater and a condenser. The still is composed of a series of chambers from 12 to 24, the internal construction of which is shown in Fig. 29. Each chamber consists of a peculiarly perforated plate *A*, a drop pipe *B*, a seal *C*, into which the drop pipe from the plate above projects, and a central standard *D*.

Returning now to Fig. 28, at the bottom of the column is a manifold *E*, with pipes *F* and *G* whereby either exhaust or live steam may be admitted. *H* designates the discharge or slop valve, controlled by a float *I* whereby a constant level of slop or spent wash is kept in the bottom chamber.

To the right of the column is seen the slop tester *J* and hydrometer *L*, whereby the spent wash may be tested to see if the spirit is being properly extracted. The steam pressure is indicated by means of a float *N* contained within a vessel *M*; a tally weight moving against a scale *K*, showing the pressure of steam entering through pipe *O* and acting against water contained in vessel *M*. Each chamber is provided with a manhole plate *P*, and a try-cock *Q*, whereby the operation of each chamber may be tested. *R* is a gage glass to show the level of the slop in the bottom chamber.

At the top of the column are three rectifying chambers fitted with boiling pipes and traps *T*, which distribute the ascending vapor and boil out the low wines returned from the wash-heater or fore-warmer.

The heater consists of a shell enclosing a series of tubes extending into an upper and lower chamber. The wash or " beer," is pumped into the lower chamber of the heater, and passes upward through the tubes to the upper chamber from which is it carried by a pipe to the plate *A* next below the rectifying plates.

The vapor from the column passes into the middle compartment of the heater and surrounds the beer tubes. The vapors give their heat to the beer and are thus cooled, the low wines being condensed and flowing back onto the uppermost rectifying plate, while the highly vaporized portions pass out to the condenser. This is of the

Fig. 30.—Detail of Perforated Plate *A*.

same general construction as the heater, the vapor being cooled and condensed to liquid by the tubes through which a constant current of cool water is passed. This enters at *U* and passes out at *V*. These tubular condensers are particularly good as they may be easily cleaned. From the condenser the spirit passes to a discharge box *W*. A portion of the flow passes into a test tube *X*, provided with a hydrometer. A trap *Y* and an air pipe *Z* provide means for the escape of gas.

As before stated, the form of perforations in the plates of a column through which the vapor pass upward through the beer or wash is particularly important. The steam must be thoroughly diffused through the beer, or else particles of mash are carried up, accumulate around the perforations, baking there and clogging them up. The clogging

Fig. 31.—Detail of Perforated Plate *A*.

and eventual stoppage of the perforations prevent the agitation of the mash carried on the plate, and a layer of mash accumulates and bakes on the head, or plate, above. Thus the operating capacity of the still is reduced and a larger quantity and greater pressure of steam is necessary with consequent waste of fuel.

It is necessary then that the form of perforation or trap through which the vapor ascends should

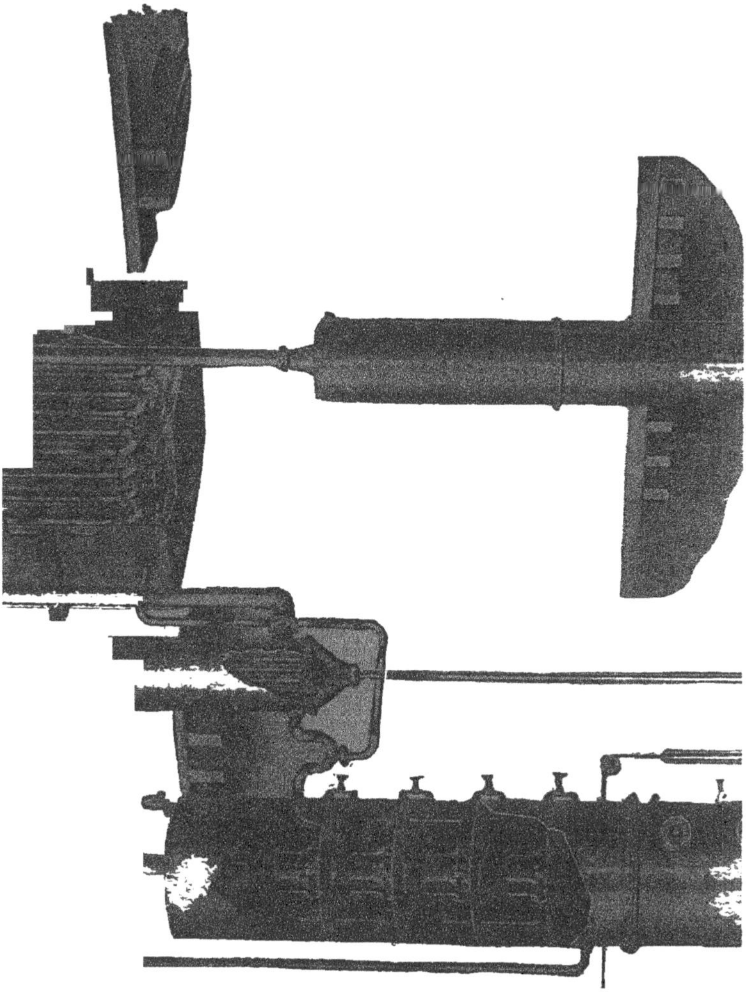

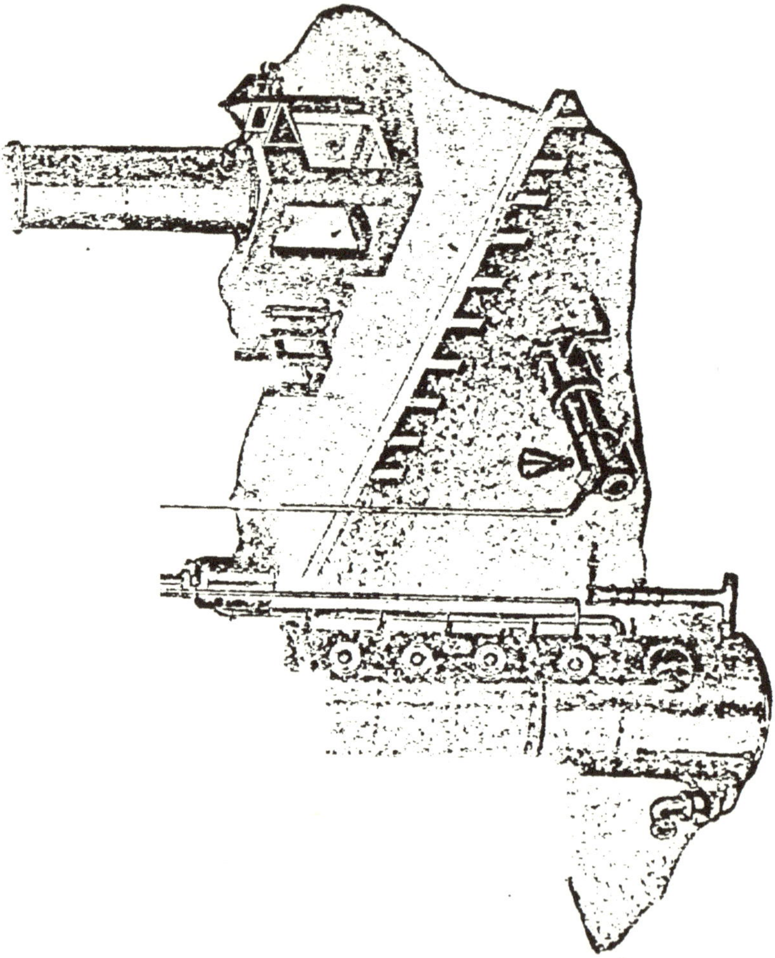

be such that agitation of the beer shall be enforced in its movement across the plate, and that the steam shall be thoroughly diffused through the beer. In the Vulcan still above referred to, these results are accomplished by forming each perforation with a tongue, as shown in the fragmentary view of a plate, Figs. 30 and 31, the tongues of all the holes being directed towards the periphery of the plate. It is claimed that by this construction the steam is diverted forward and injected into the beer, throwing the beer into vigorous motion, completely diffusing the steam and accelerating the motion of the beer from the seal *C* to the drop pipe *B*.

Fig. 32 illustrates another form of distilling apparatus manufactured by the same company, which is practically the same as the apparatus previously described except that it is provided with a "goose-necked" separator, interposed between the wash-heater and the enclosure. This consists of a series of convoluted tubes contained in a tank of cold water. The vapor from the heater passes into these convolutions. The heavier vapors are condensed therein and returned to the heater from which they descend into the column while the more volatilized vapors pass over into the final condenser. The *U*-bends at the bottoms of each convolution act like so many low wine chambers in the still shown in Fig. 9 the highly heated vapor continually bubbling through the condensed vapor in the *U* bend and there becoming greatly enriched and concentrated.

This apparatus, it is claimed, is applicable to the distillation of grain, molasses or cane juice and will yield 170 or 180 per cent., or the equiva-l nt to 85-90 G. L. or 34-36 Cartier.

A distinctly modern type of still though akin to the still shown in Figs. 19 and 20, is the inclined column of Gillaume, shown in section in Fig. 33

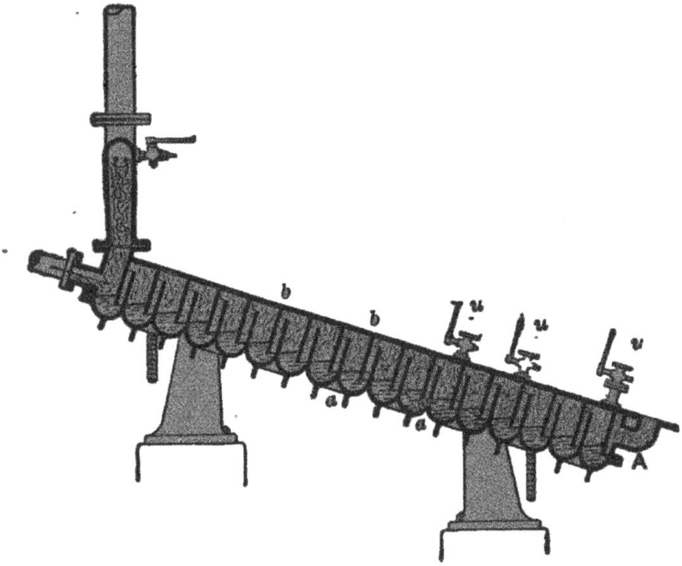

Fig. 33.—Section of Gillaume's Inclined Column Still.

and in full view in Fig. 34. Gillaume in devising this form of apparatus had particularly in mind the distillation of thick washes, and the necessity of compelling a circulation of the wash.

The bottom of the inclined column *A* is divided by lateral extending, upwardly projecting plates

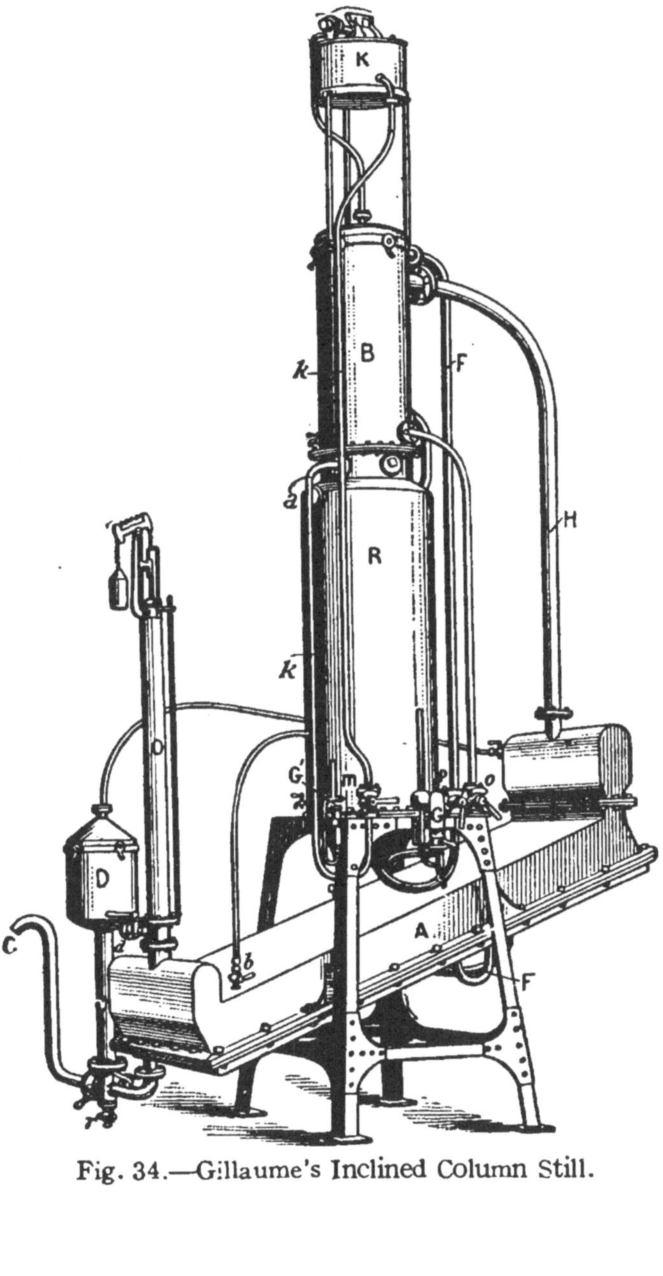

Fig. 34.—Gillaume's Inclined Column Still.

or partitions *a* forming a continuous channel through which the wash passes from side to side and from top to bottom and then out through a regulator. The upper plate of the column has downwardly projecting partitions *b* which with the partitions *a* form a series of traps. The steam enters at the bottom of the column into a reservoir, and in order to pass upward is forced beneath each partition *b* and through the washer contained in the channels of the bottom. When it reaches the upper end of the column it has passed through a continuous series of wash-filled compartments containing a constantly moving current of wash.

The vapors from the top of the column pass off to the wash heater or to a concentrator.

In Fig. 34 is shown a form of Gillaume still designed to distill all sorts of liquids whether thin or thick. The wash is supplied from an overhead tank to a regulating tank *K* from which a pipe *k* leads to a regulating tap *m*. The wash re-ascends into the wash heater *B* and when heated descends by pipe *F* into the uppermost compartments of the column *A* The vapor passes to the condenser *B*, by a pipe *H*, and the spent wash is discharged by a siphon *C*. In addition to the parts above referred to, *a* designates entrance of wash into heater, *b* exhaust test tube, *d* steam entrance tap *G* alcohol test glass, *G'* exhaust test glass, *o* valve for regulating strength of spirit, *O* steam regulator, *p* water entrance tap, *r* exit tap, and *D* the spent wash extractor.

The Gillaume apparatus is particularly valuable for the production of industrial or agricultural alcohol. It is claimed that it is easily understood and operated even by unskilled labor, while it produces a large proportion of alcohol of a high strength.

A view of a complete apparatus on a large scale is shown in the Fig. 40 in the chapter on rectification.

CHAPTER V.

RECTIFICATION.

The product of the distillation of alcoholic liquors, which is termed *low wine*, does not usually contain alcohol in sufficient quantity to admit of its being employed for direct consumption. Besides this it always contains substances which have the property of distilling over with the spirit, although their boiling points, when in the pure state, are much higher than that of alcohol. These are all classed under the generic title of *fusel-oil*; owing to their very disagreeable taste and smell, their presence in spirit is extremely objectionable. In order to remove them, the rough products of distillation are submitted to a further process of concentration and purification. Besides fusel-oil, they contain other substances, such as aldehyde, various ethers, etc., the boiling points of which are lower than that of alcohol; these must also be removed, as they impart to the spirit a fiery taste. The whole process is termed *rectification*, and is carried on in a distillatory apparatus.

As before stated, the wash as discharged into the still consists of alcohol mixed with water and a variety of impurities from which the alcohol must be separated. In order that the process may be better understood we will assume that a mixture of pure alcohol and water is to be operated on in place of the wash as above referred to. Distillation in this case is intended to deprive the water of its alcohol, the operation theoretically leaving water in one chamber and alcohol in another. This is accomplished by reason of the differences in the boiling points of water and alcohol. The alcohol vaporizes at a lower degree (173° F.) than water (212° F.) Thus the liquid at the end of the operation has been divided into two parts or *fractions*.

This, however, is not a clean division for the reason that while in the beginning the vapors contain a large quantity of the more volatile alcohol, at the end they will contain a large portion of the less volatile water. The whole of the alcohol will be separated in this manner, but it will still be mixed with some water and in order to again divide the alcohol from the water the first distillate would have to be redistilled until at last the water is reduced to a minimum or entirely eliminated, if possible.

But as it requires less heat to vaporize alcohol than water, so it also requires more cold to condense alcoholic-vapor than water-vapor. If then we pass the mixed vapors into a condensing chamber cooled to a certain temperature low enough to condense water-vapor but not the alcohol-vapor, then the water-vapor will fall down as water while the alcohol-vapor being uncondensed passes on to another chamber where its temperature falls to a point where it in turn condenses into liquid.

In intermittent distillation, as by the simple still, the vapors of mixed alcohol and water at first contain a great deal of alcohol and a little water, then more water and less alcohol, and then a great deal of water and hardly any alcohol. It may be asked: " Why not take only the runnings rich in alcohol and leave the others?" The answer to this is that if this be done then *all* the alcchol is not extracted from the wash and there is just that much loss. The solution of the problem is to get all the alcohol out mixed with the water that is inevitably with it and then redistill this result thus getting out (sifting away) some of the water, and again distill this result, and so on until only pure alcohol is left. This, however, is a very troublesome business and has been abandoned as a means of removing impurities such as water, the ethers, and fusel oil except by makers of whiskey, brandy and other beverage spirits, in favor of continuous distillation and continuous rectification.

It will be seen from what has gone before that there are two means of separating alcohol and water; one by an initial difference in heating and by a further difference in cooling or condensing.

It is on this foundation that the whole art of fractional distillation or rectification rests. While we have for illustration been considering a mixture of pure alcohol and water, the wash or liquid formed by the fermentation of grain, etc., contains a variety of ingredients of different boiling points, some more volatile than alcohol, some less. The fermented wash consists first of non-volatile or only slightly volatile matters, such as salts, proteins, glycerin, lactic acid, yeast, etc., and second, volatile bodies such as alcohol, water, various ethers, etc., fusel oils and acetic acid.

When wash is distilled in the ordinary simple or pot still, the first part to come over consists of the very volatile matters,—more volatile than alcohol even,—that is, the ethers mixed with some alcohol. This is known as the fore-shot or first runnings, and is collected separately. When the spirit coming over possesses no objectionable odor, the second stage has begun. This running would be of the alcohol proper, getting weaker and weaker, however, as the running continues and this would be caught separately as long as it is of sufficient strength. At last would come the weak spirit containing much fusel oil. It is to be understood, however, that there is no defined line between these divisions. They graduate one into the other. The first and last runnings in the old practice were mixed together and distilled with the next charge. When a strong spirit was required, rectification would be repeated several times. It is customary, however, with the improved modern apparatus, to produce at the outset spirit containing but little fusel oil and at least 80 per cent of alcohol. This is then purified and concentrated in the above manner and afterwards reduced with water to the required strength.

Another cause of the offensive flavor of the products of distillation is the presence of various acids, which exist in all fermented liquors; they are chiefly tartaric, malic, acetic, and lactic acids. The excessive action of heat upon liquors which have been distilled by an open fire has also a particularly objectionable influence upon the flavor of the products.

The first operation in the process of rectification is to neutralize the above-mentioned acids; this is effected by means of milk of lime, which is added to the liquor in quantity depending upon its acidity; the point at which the neutralization is complete is determined by the use of litmus paper. In the subsequent process of distillation, the determination of the exact moments at which to begin and to cease collecting the pure spirit is very difficult to indicate. It must be regulated by the nature of the spirits; some may be pure 20 or 30 minutes after they have attained the desired strength; and some only run pure an hour, or even more, after this point. The product should be tasted frequently, after being diluted with water, or a few drops may be poured into the palm of the hand, and after striking the hands together, it will be known by the odor whether the spirit be of good quality or not; these two means may be applied simultaneously.

The process of rectification may be carried on

in the apparatus shown in Figs. 35 and 36. *A* is a still which contains the spirit to be rectified; it should be four-fifths full. The condenser *E* and the cooler *G* are filled with water. After closing the cocks *L* and *I*, the contents of the still are heated by steam, which is introduced at first slowly. The vapors of spirit given off pass, by tubes *b*, above each plate *a*, of the series in column *B*, and escape through *C* and *D* into the condenser *E*, where they are condensed on reaching the lentils *d d'*, and return in a liquid state through pipe *f* and connections *g g'* to the upper plates of the column *B*. In these return pipes the liquid is volatilized, and constantly recharged with alcohol to be again condensed, until the water in the condenser is hot enough to permit the lighter alcoholic vapors to pass into the coil *c c c*, without being reduced to the liquid state. When this is the case, the vapors pass through *F* into the cooler *G*, where they undergo complete condensation. Great care must be taken that the heat is not so great as to permit any of the vapors to pass over uncondensed or to flow away in a hot state; and also to keep up a constant supply of water in the cooler without producing too low a temperature; the alcoholic products should run out just cold. The highly volatile constituents of the spirit come over first, that which follows becoming gradually purer until it consists of well-flavored alcohol; after this comes a product containing the essential oils. The more impure products are kept apart from

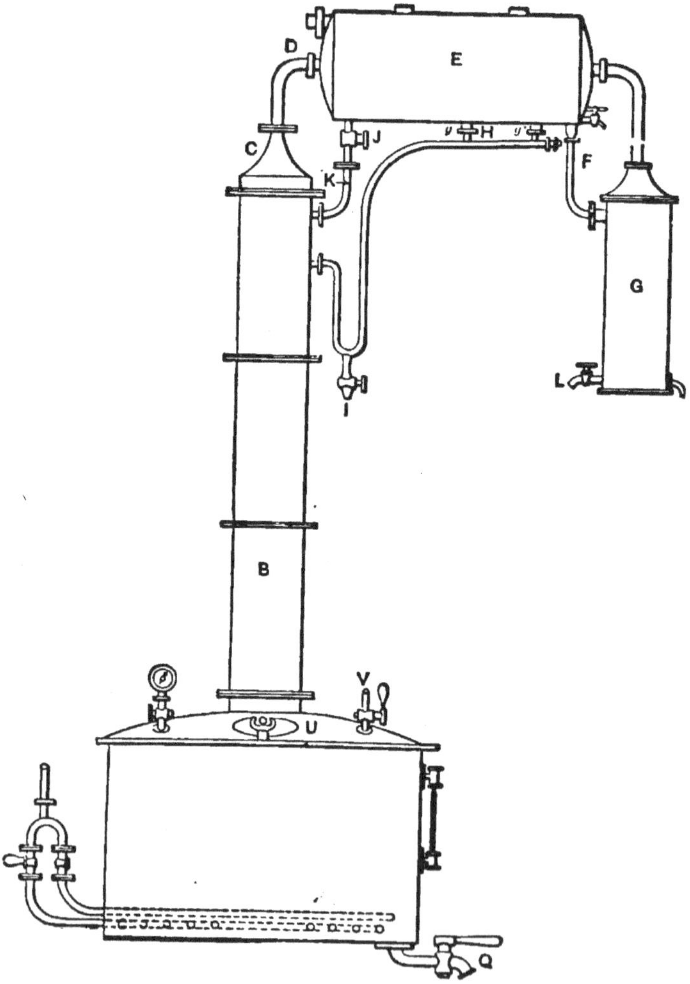

FIG. 35.—Rectifying Still.

the rest and re-distilled with the next charge. Some hours generally elapse before alcohol begins to flow from the cooler. The purest alcohol is obtained while its strength is kept between 92° and 96° Baume, and the operation is complete

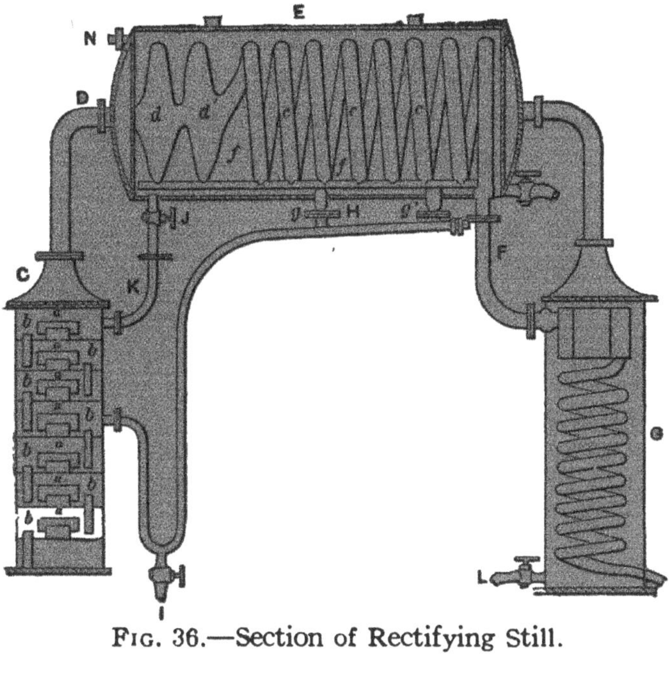

FIG. 36.—Section of Rectifying Still.

when the liquid flowing through the vessel marks not more than 3° or 4° Baume; it is better, however, to stop the still when the backing or " faints " indicate 10°, because the product after this point contains much fusel-oil, and is not worth collecting.

In order to cleanse the apparatus—which should

be performed after each working—the still *A* is emptied of water by opening the cock *Q*. The contents of the condenser are then emptied in like manner by opening the cock *J*, through which they flow upon the plates in the column *B*, and wash out essential oils which remain in them. These two cocks are then closed, and the door *U* in the still head is removed. The water in the cooler *G* is then run by means of pipe into the still *A*, so as partially to cover the steam-coil in the latter. After again securing the door *U*, a strong heat is applied, and the water in the still is well boiled, the steam evolved thoroughly cleansing all the different parts of the apparatus; this is continued for 15 or 20 minutes, when the heat is withdrawn and the still left to cool gradually.

In the intermittent rectifying still above described the impure products are distilled with the next charge. In the apparatus as perfected and used in large distilleries or rectification plants, the division of the several products composing the phlegm or raw spirit is made at one time and continuously on the principle now to be described.

It was stated in the beginning of this chapter that the various impurities in alcohol, the ethers, the water and the fusel oils, have each their own vaporizing-point and each their own condensing point. As this is so, they may be separated from each other and from the alcohol on the same principle as we have seen that water is separated from the mixture of pure alcohol and water; that is, by fractionation, as it is termed, or by "sifting out" one body from another.

Thus in fractional distillation, each condenser or retort in the apparatus shown in Fig. 37, above acts as a sieve or trap, letting pass the most volatile substances but retaining those of a less degree of volatility. By passing the mixed vapor together through a good condensing medium the temperature of which is lower than the boiling

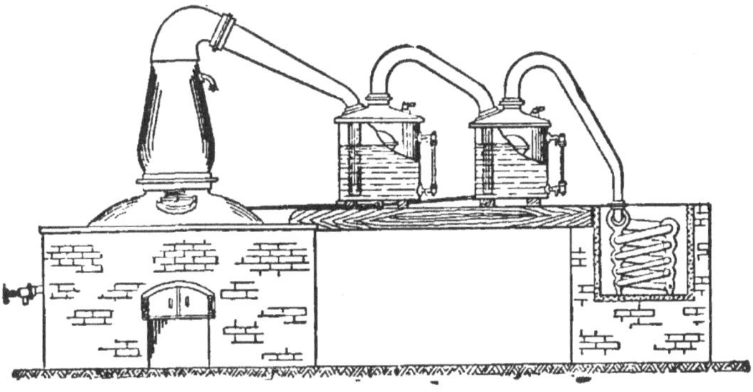

Fig. 37.—Fractional Distilling Apparatus.

point of the less volatile, but not so low as the boiling point of the more volatile the vapors of the less volatile liquid will be condensed, while the more volatile will retain their gaseous form. Thus by having a number of condensing mediums each one slightly lower in temperature than the other, the various vapors with their various points of volatilization will be successively condensed, allowing the passage of the more volatile vapors over to the condenser beyond.

If we had mixed gravel and sand and desired to separate the gravel into assorted sizes and get the sand by itself, we would pass the mass through a series of sieves of gradually smaller mesh. The first sieve of course would catch all the largest pebbles, the next in size would let all the second sized gravel through, and so on until the final sieve would have separated the coarse sand from the fine. In this figure of illustration, the coarse pebbles may be taken to represent the water and the fusel oils which are mixed and partly tend to rise with the alcohol, and the alcohol may be represented as the gravel larger than the sand, and the fine sand as the etheric vapors. If this gravel were forced upwardly through a series of sieves gradually growing finer, it would be analogous figuratively to the upward passage of the vapors through a distilling column composed of plates or chambers; the water and fusel oils would be retained in the lower portion of the column and continually sent back there; the alcohol would pass into the upper chambers of the column and the ethers or fore-shots would pass out from the very fine sieve at the top of the column.

The vertical chambers above each plate of the rectifying column are to-day used as the separate eliminating chambers referred to above. It has been found in practice that as before stated, each plate of a column contained upon it liquid of a certain temperature and above it vapors of a certain degree of vaporization. That is, in a continuous column fed regularly by condensation from above and supplied with a constant flow of phlegm, each plate carries upon it a liquid of constant composition relative to the boiling point of the fluid on that plate. As many extractions may thus be made from the various plates as there are different liquids to be isolated. Thus by tapping different portions of the column, vapors of different degrees of vaporization are found and may be carried off and the phlegm be thus fractionated. In the case of one column the first runnings or fore-shot would be found in the upper portion of the column to which they would have risen by reason of their degree of volatility. The last runnings or oils, aldehydes, etc., would be found in the lower portion of the column still mixed with the spirit, while upon the plates of the middle portion of the column would be found the vapor of the alcohol freed from the fusel oils and from the ethers.

It is understood, of course, throughout this description that the liquid being treated is not wash but phlegm; that is, the raw spirit containing the fusel oils, ethers, water and alcohol.

Fig. 38 represents a simple rectifying apparatus designed for small or medium sized plants, and manufactured by the Vulcan Copper Works Co,. of Cincinnati. The still is upright, with a chambered column above it, of the usual type. The chambers are fitted with a vapor boiling pipe and cap and a drop pipe, and each is provided with cocks whereby it may be drained for cleansing.

Above the column is a separator, comprising a casing containing a series of tubes. The vapor from the column circulates around the tubes through which passes a current of cool water. The condenser is of the same construction as the separator and is provided with a gage glass and a draw-off cock. The operation is the same as in other simple rectifiers; part of the vapor from the column is condensed in the separator and passes back on to the upper plates, while the more highly varporized portions pass over into the condenser.

The diameter of the still is large relatively to its depth so as to yield an economical and at the same time highly effective distribution of heat through the charge. This also affords an extended boiling area from which the vapor rises evenly and regularly, thus ensuring conditions peculiarly conducive to produce the best fractionating. The floor space required for this still and others of the same character built by this company is very compact and excessive weight on the top floor of the building is dispensed with.

We have shown in Figs 39 and 40 two forms of rectifying apparatus, one a twin column Barbet rectifier and the other a rectifier of the Gillaume type combined with inclined column still.

In the twin column apparatus, Fig. 39, the first column or clarifier *A* receives the raw phlegm and accomplishes the elimination of ethers. The clarified phlegm passes then to the second column

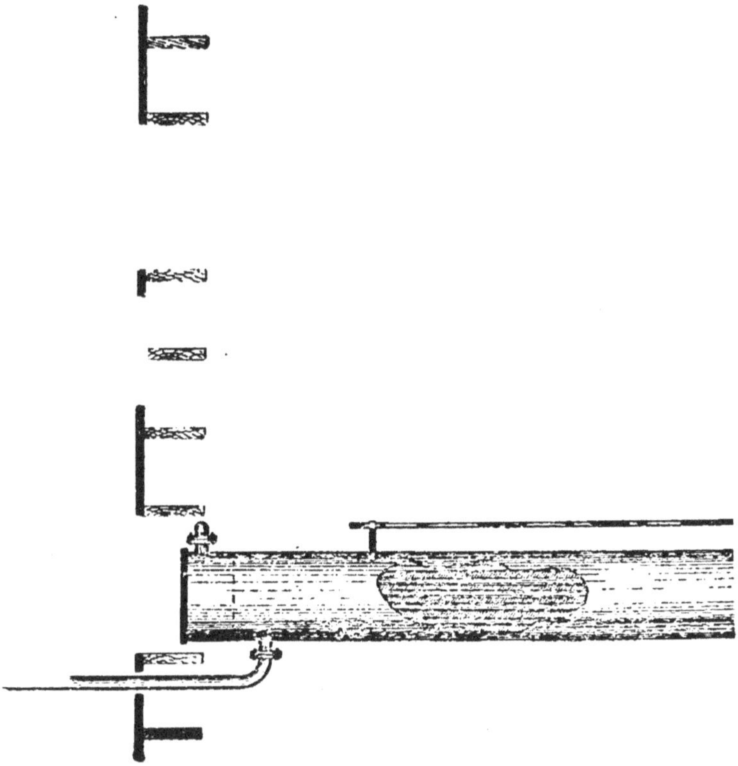

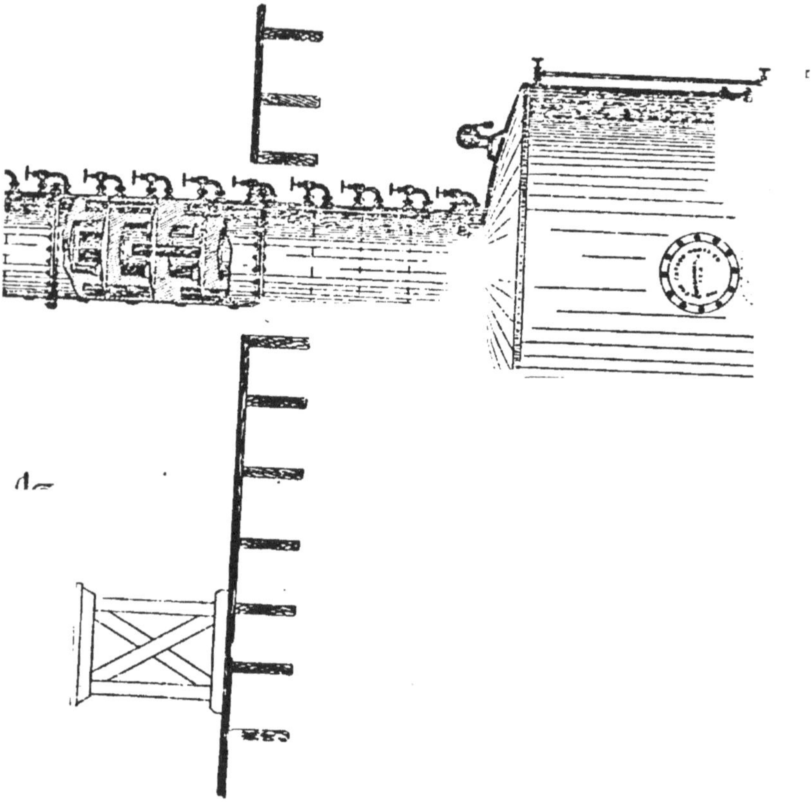

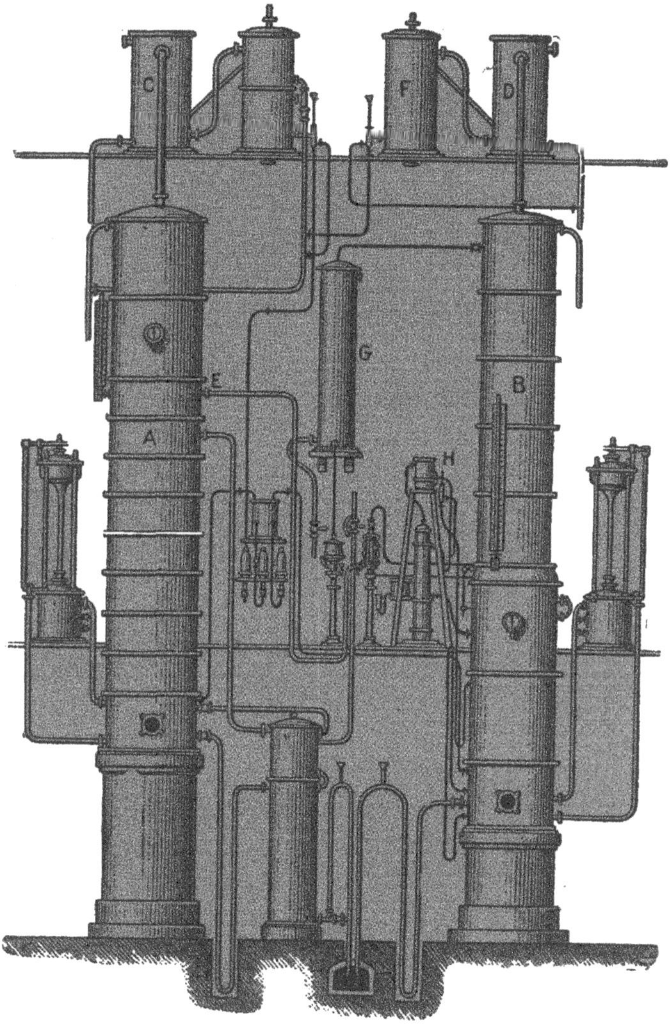

Fig.39.—Twin Column Barbet Rectifier.

where the alcohol is separated from the last runnings or fusel oil. In other words, the phlegm or impure raw alcohol is only raised to such a temperature in the first column as to drive off the very volatile constituents such as the ethers. These therefore pass off at the top of the first column into the condenser *C*, the retrogradation or condensed alcohol being returned to *A*, while the boiling phlegm taken from the middle of the column and still containing the aldehydes, oils, etc., is conducted by a pipe *E* to the second column *B* wherein the last runnings or amylic oils, etc., are separated from the purified spirits.

The vapors in this column are carried to the condensers *D* and *F* and from there to a refrigerator *G*. The fusel oils are extracted from the plates slightly below the center of the column and are carried to an oil concentrating apparatus *H*.

In the most complete forms of apparatus used to-day, there is a variation of this construction. The first runnings, middle runnings and the last runnings are each led off from the main column to separate coolers, condensers, etc., and the purified result from each of these columns is in turn led to a trunk rectifier common to all where the product is redistilled and entirely freed from impurities. This gives a very high grade of alcohol by a process practically continuous. At the same time the impurities are not returned to the first or main column to contaminate the vapors therein and add to the amount of fusel oils contained on

Fig. 40.—Gillaume's Rectifier and Inclined Still.

the lower plates. In construction of this character there is a very large saving in the cost of the fuel and the result is much better in every way.

FIG. 40.—GUILLAUME'S DIRECT DISTILLATION-RECTIFICATION APPARATUS FOR "AGRICULTURAL" DISTILLATIONS.

A Distilling Column.
a Tank for Wash to be Distilled.
b Cold Water Tank.
C Rectification Column.
D Final Purification Column.
I Wash Heater.
K Condenser.
K' Refrigerator of Ethers.
O Refrigerator for high-grade Alcohol and the First Runnings.
Q Refrigerator for the Products of the Last Runnings.
R Spent Wash Extractor.
r Siphon Carrying off Spent Wash.
S Steam Regulator.
s Tap and Pipe for Carrying Wash to Distilling Column.
U Water Regulator.
u Taps for the Extraction of Intermediate Impurities.
V Receiver Accumulator.
v Tap for the Extraction of the Last Runnings.
X Test Glass for the High-Grade Alcohol.

Y Test Glass for First Runnings.
Y' Test Glass for Last Runnings.
Z Test Glass for Determining Degree of Exhaustion of Spent Wash.

In this apparatus the still proper is of the form heretofore described on page 78. The liquid to be distilled enters at the top of the inclined column *A* and descends to the base thereof. The alcoholic vapor rises through the column and passes off from the head thereof into the rectifying column. At the head of the column *A* it has a strength of 40 to 50 per cent. The column *C* is supported upon an accumulating reservoir *V* which acts to regulate the flow of the phlegm through the rectifying column and prevents too great an exhaustion of the plates of the column. It acts as a reservoir to contain any excess of phlegm or to supply an additional amount of phlegm to the plates when they have become nearly exhausted.

The oils or products of the last running accumulate at the base of the column, and are carried off to their special refrigerator *Q*. The alcoholic vapors concentrate while rising in the column and quickly attain a strength of 92 to 94%. At a height within the column corresponding to the plates whereon alcohol of that strength is to be found, there are provided three taps *u* whereby the middle runnings or medium grade of alcohol may be drawn off, which have a maximum concentration of 92 to 94 per cent. Above these middle plates the alcohol vapors are completely separated from the products of the "tail" that is the aldehydes, amylic oils, etc., and at the upper portion of the column there is found the condenser *K* which separates the products of the head; that is the first runnings from the alcohol which has passed over with such products to the condenser. The alcohol so separated is completely rectified in the column of final purification *D* and the finished alcohol is cooled in the refrigerator *O* below the column of final condensation. In this apparatus the gauge glasses which regulate the exit of the various alcohols and mixtures are controlled by taps having verniers or scales whereby they may be very carefully adjusted, to regulate the relative proportion of the various products. This apparatus is able to produce about 75 or 80 per cent. of first-class alcohol, 10 to 15 per cent. of middle class alcohol, and 5 per cent. of ethers and 5 per cent. of fusel oils, the alcohol produced being about 96 per cent.

The alcohol is thus obtained in one single operation and with, it is asserted, only a very small loss in rectification. The apparatus is claimed to be so simple that it may be operated even by unskilled farm labor. It is also claimed that purification by chemical treatment or filtration is unnecessary with the Guillaume apparatus. It may be stated, however, that the Guillaume system has many opponents.

The capacity of the rectifying apparatus has a good deal of influence upon both the quantity and the quality of the spirit obtained. Besides being much more difficult to manage, a small apparatus will not yield so large a proportion of spirit as a more capacious one, nor will its products be of equally good flavor. The proportion of alcohol which may be obtained from a successful rectification is very variable; it depends upon the nature of the spirit rectified, the method of extracting the sugar, and the manner of conducting the distillation; it will also be in inverse proportion to the quantity of fusel-oil contained in the raw spirit. The average loss of pure alcohol during the process of rectification is generally estimated at about five per cent.

In addition to the rectifying as above described, alcohol may be further purified by filtration through charcoal, by chemical means or by electrolysis. The last two methods have not so far been successful. The chemicals used merely act to disguise the disagreeable taste or smell of the spirit and do not really purify. They but substitute one impurity for another. The agents used are many—sulphuric and nitrate acids, soaps, oils and fats soda, lime and potash have each and all been tried, but with no permanent success. As agents for disguising the taste of new and raw spirits, alcoholic extracts of fruits have also been used.

Purification and aging by electricity has been tried many times and in many different forms, but so far has not been commercially practicable.

Filtration still remains the best and simplest adjunct to the rectifier. In small plants, a filter bed several feet in thickness of bone black or beachwood or charcoal is used, laid upon a foundation of gravel in a filtering tank. In the larger plants a series of these vats is used, the charcoal being used in lumps varying from ¼ to ½ inch in diameter. Two different views of the purification by charcoal are held—one that the charcoal purifies by chemical means, the other that it is purely a physical filtration agent. After filtration the charcoal must be steamed to recover the spirits retained therein and should be heated to a red heat every now and then to cleanse and regenerate it.

CHAPTER VI.

MALTING.

Wheat, oats, rye, potatoes, and other amalyceous or starchy materials contain starch insoluble in water and to render it soluble, and to change the starch to maltose they must be mashed with a certain small proportion of malt,—or grain in which germination has been artificially induced and then interrupted at a certain stage. This increases the diastase contained in the grain so germinated, and this diastase is able to transform starch into soluble form. Hence, malted grain gives lightness and liquidity to the wash, and prevents the starch falling to the bottom of the mash tub or " back," and also prevents the starch falling to the bottom of the still and consequent burning.

While all varieties of grain including rice are suitable for the preparation of malt, barley is preferred to all others, and is most commonly used.

The best barley for malting is that having the following characteristics; a thin skin; a mealy interior; grains of a uniform size; of the greatest weight; which has been stored for three months. Barley on harvesting has but slight germinating power. The reason for the uniformity in the grains lies in necessity of a uniform steeping of the grain so that the period of germination shall be the same for the whole mass.

Like all materials for distillation, the barley should be thoroughly cleaned of impurities—not only dust, seeds and weeds, but fungi and bacteria. This may be partly accomplished in the ordinary fanning mills, usual on farms, but a better machine would be a "tumbling box" of wire mesh. This is inclined, so that grain put in the upper end, will pass downward to the lower, being thrown about as the box or cylinder is rotated. The dust, seeds, etc., fall through the meshes of the wire as do the smaller grains. After this cleaning, the barley should be thoroughly washed. This may be either done in the steeping vat itself—and the water afterwards drawn off—or in special machines. If the barley be allowed to soak in water for a day or two, the later washing will completely cleanse it. This preliminary cleaning is most important as impurities reduce the germinating power of the grain, as well as introduce bacteria inimical to fermentation.

Washing in some instances is done by forcing compressed air into the steeping tub, thus violently agitating and swirling the water therein, and washing away the impurities. Another method is by passing the steeped grains along a trough supplied with moving water, the trough being provided with rotary agitators. Any fairly ingenious mechanic could devise a capable cleansing machine. Care being taken that it shall not injure the grains.

After cleansing, the barley should be steeped. For this purpose tanks of metal or cement are to be preferred to wood. All vats should be kept thoroughly cleaned by frequent scrubbing with lime water.

The barley placed therein should at all times be entirely covered with fresh water to a depth of a few inches, and for the first few hours the grains should be carefully stirred in order that no grain should escape wetting. At the end of that time the still floating grains should be removed.

In 36 or 48 hours the grain will usually be sufficiently steeped,—but this varies with weather conditions. The warmer the water the quicker the steeping, and in winter proper steeping may not be accomplished before four or five days.

A simple test is to rub the grain strongly between the hands, If it is entirely crushed, and no solid matter is left it has been steeped sufficiently. Barley should be capable of compression lengthwise and the hull should become easily detached. It should be easily bitten, and not crack under the teeth. In order to prevent fermentation in summer, it is well to renew the water a few times during steeping. Over steeping is worse than under steeping.

After the barley is in proper condition the vat or tank is opened and the water drained away.

The draining should be complete, and therefore the grain should be left to drain about 12 hours.

Germinating. The grain is now taken to the malting floor. In practice it is well to locate the steeping vat above the malting floor, so that the steeped grain may be run down on to the floor without inconvenience.

It is best to first spread the grains out on the floor to a depth of a few inches in order that it may somewhat dry out. This is not necessary when it has not been steeped to a great extent.

After 10 or 12 hours of drying, the grain is placed in a heap until warm to the touch, which may occur in from 12 to 24 hours. It is then disposed in a layer from eight inches to 20 inches thick. This is called the "wet couch." The lower the temperature the thicker the couch should be. It should be turned every six or eight hours in this stage.

The heat so germinated after 25 or 30 hours produces at the end of each grain a small white rootlet. The grain in the middle of the layer is the first to sprout, as it is the warmest, hence the couched grain should be frequently turned so as to give all the grains a uniform heat, and a uniform germination. At this period the grains beneath the surface are dampish to the touch.

The height of the couch is now successively lessened to layers of from six to two inches called "floors," the height of each floor of course depending on the temperature, as before.

It is to be understood that the growing grain requires both dampness and air, hence the " floor " should not be thinned so rapidly as to deprive it of moisture, and the barley should be turned at least twice a day to give each grain a proper aëration. During this period the small white rootlets or radicals should be white and shiny. If they begin to fade, it is a sign that they lack water and the grain should be sprinkled. Germination usually requires from a week to ten days, or sometimes two weeks, depending on the previous steeping, the quality of the grain and the temperature. When the fibers or rootlets of the grain are about equal to the length of the grain, germination is complete.

It used to be considered that malt was in its best condition in eight or ten days. To-day, however, " long malt " is used,—requiring a germinating period of twenty days, being frequently moistened and turned during this time, and the temperature being kept at 65°F. This malt is very strong in diastase.

The effect of germination is to produce a change particularly favorable to mashing. The barley becomes sweetish, the gluten is partially destroyed and what is left is soluble. Thus the fecula or starch is set at liberty and free to be acted on by the yeast used in fermenting.

March is the best month in which to malt; and while the malt is best used immediately, it can not be kept in its green state and must be therefore dried for future use.

Drying. This is accomplished either in the air of a warm, dry room in hot weather, or by means of a drying kiln. In the first process the malt is spread in a thin layer and frequently turned. In the second the grain is spread out in a layer from eight inches to a foot thick on the grain floor of the kiln.

Beneath the grain floor a fire is maintained. In the beginning the temperature of the drying floor should be about 85° F., but this is increased gradually to about 104° F. until most of the moisture has been removed. The heat is then raised to from 120° F. to 130° F., thus completely drying the grain.

The germinated green or dried barley is called malt. It is of good quality when the grain is round and flowery; when it crumbles easily and when its taste is sweetish and agreeable. Pale malt or that which has been hardly altered from its original color is the best for distillation.

Before the malt can be used it should be screened so as to remove the rootlets.

Two hundred and twenty lbs. of barley should yield from 275 to 350 lbs. of green malt, about 200 lbs. of air dried malt, and from 175 to 190 lbs. of kiln dried malt.

In large plants malting is now so carried on that the steeping germination and drying are all accomplished in one vessel or container, by one continuous operation. This vessel is commonly in the form of a drum of sheet iron, revolving at a very slow speed. Moist air is introduced and the carbonic acid laden air withdrawn. After germination the malt is dried by passing in dry air at the proper temperature.

As these systems are only adopted to large distilleries, using expensive machinery, further reference to them is not considered necessary in this volume.

Previous to use the malt must be finely ground or crushed either before or after mixing with the materials to be mashed. It is not necessary or advisable that the malt be reduced to flour. The use of malt with other materials in order to form a fermentible mash, will be considered in the chapters on specific mashes.

CHAPTER VII.

Alcohol from Potatoes.

In certain countries, as for instance Germany and France, potatoes form the greatest source of alcohol, particularly for industrial purposes. With the possible exception of corn and beets they will probably be most used in America.

The best potatoes for distilling are those which are most farinaceous when boiled. In other words, those which are "mealy" and most appetizing. These give the largest yield of alcohol per bushel. The best season of the year in which to use potatoes is from October to March, when they germinate.

The potatoes should be kept in dry cellars, and at even temperatures, warm enough to prevent freezing and yet not so warm that they will rot or sprout. Diseased potatoes may however be used, if they have not been attacked by dry rot, though they are not so easily worked. Frosted potatoes may be also used, but they must not have been completely frozen.

Before being steamed, the potatoes should be washed, either by hand or by a machine, care being taken to remove all stones, clods of earth, and other foreign substances which might impede the subsequent operations.

There are three main methods of saccharifying the fecula or starch of the potato. The first and most important by reducing the tubers to a pulp, and malting the entire mass. The second and third, by rasping the potatoes and so separating the fecula or starch grains from the mass, and then making a thin liquor or wash containing this fecula.

Originally, in the first process, the washed potatoes were submitted to the action of boiling water, but later cooking by steam at a temperature of 212° F. was used, as being much more convenient to handle and more effective in action. The object of steaming is to break the coating and reduce the contents thereof to a pasty condition, wherein the starch is more effectively acted on by the malt and yeast. Ordinary steaming does not, however, render the pulp sufficiently pasty; some of the starch remains undissolved and is lost, hence in the modern practice, steam is turned into the steaming vat under a pressure of three or four atmosphere (45 to 60 lbs. to the square inch).

High pressure steaming will be later described but the simple and older method of mashing and apparatus therefor, used prior to 1870, was as follows:

Fig. 41 shows a section of a steaming vat. This consists of a conical wooden tub *H* provided at its top with a suitable cover *O* having a trap or door *P* for putting in the potatoes. This as shown, consists of a hinged lid, having a button *p* or other fastening means. This lid and cover should be of course steam tight, and it would be better to have it clamped down by a screw clamp than held by a button.

Somewhat above the bottom of the vat, a steam inlet pipe *I* enters, connected at its other end by a coupling *i* with a suitable steam generator (see Fig. 43). Preferably the outlet of this pipe is

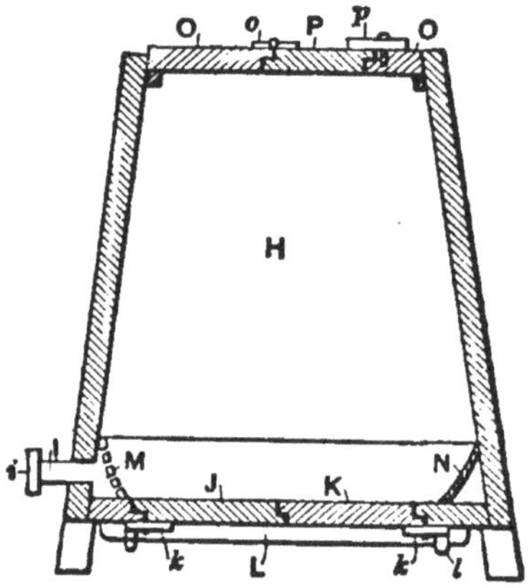

FIG. 41.—Steaming Vat for Potatoes.

screened by a perforated plate *M* so that it may not be clogged by the pulp. It is also best that a filling piece be placed at the junction of the bottom with the sides in order that there be no sharp corner from which the pulp may not be easily cleaned out.

The bottom of the vat may either have a discharge door at the side as in Fig. 44 or at the bottom, as in Fig. 41.

An under side view of the latter construction is shown in Fig. 42. The bottom of the vat is made in two parts or doors *J K*. These are held closed by a transverse bar *L* inserted at its end into a stirrup l^1 and supported at its other end by a button *l*, or other means.

While various forms of steam generators may be used, Fig. 43 shows a simple construction well adapted to the needs of a small distillery. *D* desig-

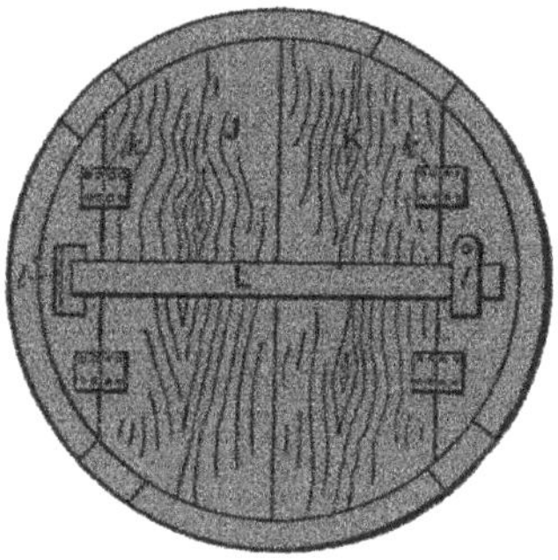

FIG. 42.—Bottom of Steaming Vat.

nates the brick work of a furnace, and *A* the boiler. This is so set that an annular space *E* surrounds the sides of the boiler, through which the products of combustion must pass.

The head of the boiler is connected by a pipe *B* and collar *b* to the steam inlet pipe *I* of the steaming vat, heretofore described, as by the collars *b i*.

A filling tube *C* enters the boiler and project nearly to the bottom, and the water outlet-pipe *ι* with cock *f* leads off from the upper water line. The tube *C* forms also a safety valve, for if the steam pressure becomes too great in the boiler and connected vat, it will force water up and out through the tube. If, however, the water falls below the level of the lower end of the tube, steam will issue and warn the attendant that water is too low. It would be best however, to provide a steam gauge, whereby the pressure of

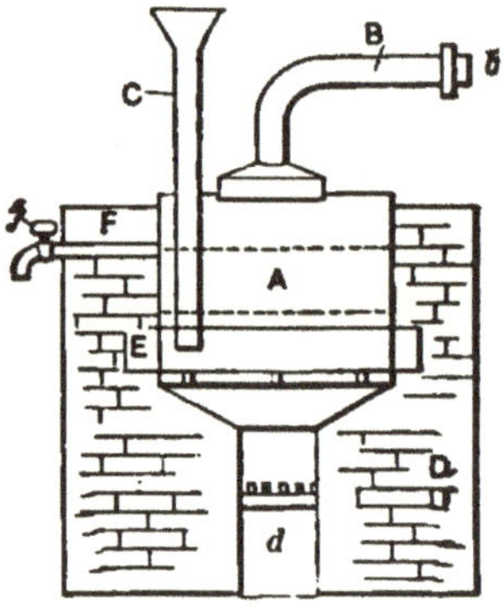

Fig. 43.—Steam Generator.

steam in the boiler and vat could be accurately indicated.

It is to be noted that when steamed the potatoes will swell and occupy more space and that the steam vat should therefore not be much more than two-thirds filled with potatoes.

With the steaming vat above shown, the potatoes are delivered mixed with a considerable quantity of water, but a better plan is to have a perforated false bottom to the tub, whereby the condensed water may be carried away, the steamed potatoes remaining behind.

Two hours of steaming should reduce the potatoes to proper condition, which may be tested by introducing a pointed iron rod through a suitable aperture, normally kept closed. If the rod passes freely inward, the potatoes are done and may be discharged into the crusher, shown in Fig. 44. In this Fig. the steaming vat *A* is shown mounted above the crusher. A pipe *B* with cock *b* leads to the steam generator. The steamed potatoes are shoveled out through the door *a*, which is usually held closed by means of the clamps or buttons *a′ a″*.

The crusher consists of a hopper *C* whose bottom fits closely against two adjacent smooth faced rolls *H I* of iron. These are driven by gears *D E*. The shafts of these gears have cranks *d d* whereby it may be operated. These gears are unequal so that the rolls shall move at different speeds, and thus one will have a grinding action against the face of the other. A counter weighted scraper *e* bears against the face of the roll.

The crushed potato pulp passes between the rolls and into a bin beneath, having adjustable walls made of boards *F*, sliding in suitable guides *f*, from which the pulp may be shoveled into the mashing tank or " back." The crusher might, however, be arranged to deliver immediately into

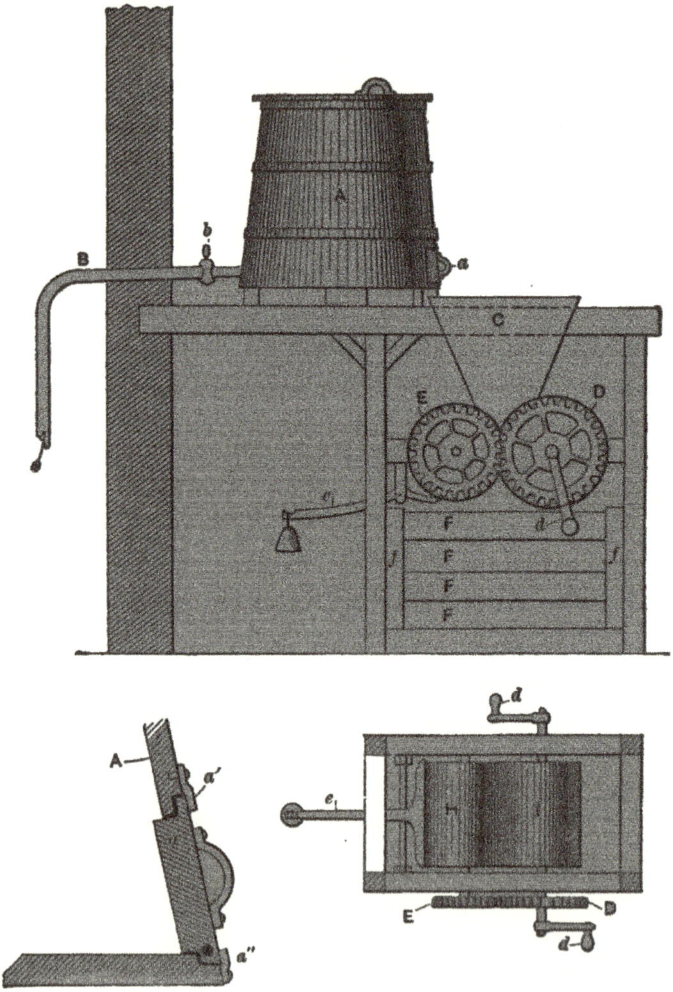

Fig. 44.—Potato Steamer and Crusher.

the mashing tank, if the latter is provided with means for stirring the delivered pulp.

The pulp or paste thus made is now placed in a vat, holding about 650 to 850 gals.. in which the saccharification takes place About 2200 lbs. of the crushed potatoes and 155 lbs of broken malt are introduced, and immediately afterwards water is run in at a temperature of about 97° F. to 104° F., the contents being well stirred with a fork meanwhile. The vat is then carefully closed for half an hour, after which boiling water is added until the temperature reaches 140° F., when the whole is left for three or four hours. The process of fermentation is conducted in the same vat. Alternate doses of cold and boiling water are run in upon the mixture, until the quantity is made up to 700 or 775 gallons, according to the size of the vat, and so as finally to bring the temperature to 75° F. or 79° F. Five and a half to six gallons of liquid brewer's yeast are then added, and fermentation speedily sets in. This process complete, the fermented pulp is distilled in the apparatus devised by Cellier-Blumenthal (see Fig. 15) for distilling materials of a pasty nature; the product has a very unpleasant odor and taste.

The process above described is the old method of pulping the potatoes by using steam. Under the modern method, however, and with modern apparatus, in preparing potatoes for distillation in large quantities, the steaming of the material is accomplished at one time and under a high steam pressure. The apparatus is also used for the preparation of corn, potatoes and other starch-containing substances.

There are many apparatuses which have been devised for the purpose, but the principle on which they work is practically the same in all cases. They comprise a closed tank, fitted with stirrers, agitators, or other means for mixing and

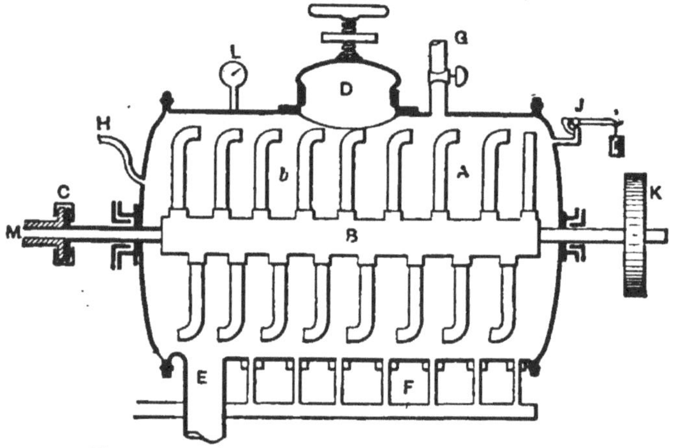

FIG. 45.—Bohn's Steamer and Crusher.

comminuting the contents, means for admitting steam under pressure, means for cooling the mixture to the proper mashing temperatures, and means for forcing the steamed material out of the tank.

The Steamer. One of the earliest forms of steamer was that of Hallefreund devised in 1871, and adapted for working on a large scale. A modified form of the apparatus known as Bohn's steamer and masher is illustrated in Fig. 45. This comprises a steaming cylinder *A*, having a securely closed opening *D* for the introduction of the potatoes. Centrally through the cylinder passes a hollow shaft *B*, which is rotated by the power pulley *K*. Hollow arms *b* project radially from the shaft *B*. These act as mixers of the mash and as coolers. The shaft *B* at one end is connected to a cold water supply pipe *M* as by a coupling *C*, the supply pipe being provided with a cock. *E* designates a discharge opening for the mash. A pipe *F* provides for the entrance of steam into the cylinder. *G* is a pipe through which malt is put in to be mixed with the pulp. *L* is a steam gauge and *J* a safety valve. *H* designates a water pipe. For the relation of the steamer to other apparatus, see Fig. 1.

In operation the potatoes are placed in the cylinder *A* and submitted to the action of steam at about 46 lbs. to the square inch, and at a temperature of from 266° F. to 275° F.

When disintegrated, the steam is blown off, and the potatoes crushed by rotating the stirring shaft. As the pulp must be reduced from 275° F. to 149° F., the mashing temperature, cold water is forced into the stirrer which chills the blades and quickly cools the mass.

In the vacuum mash cooker shown in Fig. 1, the steaming cylinder is partly filled with hot water at 140° F. to 150° F. The potatoes to be mashed are fed into the cylinder whole. The steamer is then closed and steam admitted while the mash is stirred until a pressure of 65 pounds is reached, when the dissolution of the starch is complete. The steam is then exhausted and the temperature reduced to 212° F. To reduce this temperature to the proper saccharifying point of 145° F., the hot air is exhausted.

Barley malt meal in the proportion of 6 to 10 per cent. is used. This has been previously mixed with cold water in the small grain masher. The malt is admitted to the cylinder and thoroughly mixed with the potato, when the mixture is withdrawn into a drop tub, where it is still further stirred. It is then cooled as described on page 15 and then fermented.

While the crushed potatoes are being cooled and stirred, a mixture of green malt with water is prepared in an adjacent vat, and when the pulp in the cylinder has been reduced to 149° F. the malt mixture is introduced into the cylinder through the pipe *G*, and thoroughly mixed with the crushed potatoes. The mass is now left to saccharify; the stirrer being operated at intervals throughout this period. This machine might be readily modified so that the steam should enter through the stirrers, by tubes attached to the arms, then the steam may be shut off and cold water sent into the arms themselves to cool the mash.

A variety of steamer used in various forms and modifications in all the larger distilleries, is known as the Henze steamer, Fig. 2. In this, there are no stirrers. The cylinder is conical, and has steam pipes leading to the interior. At the end of its cone-shaped bottom it terminates in a blow-off tube, having in it a grate formed of sharp-edged bars. In operation, steam is introduced at a pressure of one to two atmospheres until the potatoes are cooked. More steam is then suddenly admitted at high pressure and the softened potatoes forced through the grating at the bottom and into the mashing apparatus in a finely divided state.

In steaming under pressure it is best that the safety valve be so regulated that the steam will constantly blow off as this action keeps the potatoes in motion and facilitates disintegration. Care should also be taken to see that everything about the apparatus is in good condition, as in working under the high pressures used in the last apparatus there is liability of explosion. Rust should be particularly guarded against.

With this apparatus a preparatory mash vat is used into which the contents of the steamers are blown out, malt and water to form milk having been previously let into the mash vat. Blowing out is accomplished in 45 or 50 minutes at 130° F. and about one-sixth of the charge in the steamer is retained in the steamer. The mash in the vat is stirred and cooled and the remainder of the mash blown in raising the temperature to 145° F. when the mash is left to stand from half an hour to an hour. With heavy mashes, rich in sugar, even higher temperatures than 145° F. can be used for saccharifying.

The processes of crushing and saccharifying, above referred to, which are almost entirely used to-day, require steam. The following methods provide for the isolation of the fecula or starch, without steam and the production of a wash of a more watery consistency, therefore easier to handle in ordinary stills, and with less liability to burn.

Two operations are necessary by this method: First, rasping, or reducing the potatoes to a finely crushed and pulpy condition by means of a machine described in the chapter on Beet Mashing; and second, the separation of the fecula.

To this latter end the potato pulp is placed on a sieve, having side walls and net work of horse-hair, which is placed over a suitable tub. Water is run gradually through the pulp and sieve, while the pulp is rubbed up by hand. When the water comes through clear, then all the fecula of the pulp has been washed out, and the refuse left in the sieve can be thrown aside or used as a food for cattle.

For a mashing tub of say about 32 bushels capacity, the fecula from about 800 lbs. of potatoes is used. This is deposited in the mash tub with sufficient cold water to form a fairly clear paste. About twice as much water as fecula will bring the paste to proper consistency. This mixture should be constantly stirred as otherwise the fecula will sink to the bottom. About 40 gallons of boiling water are then added gradually. The mixture has at first a milky appearance, but at the last becomes entirely clear.

This liquid is mashed with about 45 lbs. of malted barley or Indian corn, ground into coarse flour. In ten minutes the mixture will be completely fluidified. It is then left to subside for three or four hours when it will have acquired a sweetish taste and be what is termed as " sweet mash." The fluid is then further diluted by the addition of sufficient water to give about 290 gallons of wash. Two or three pints of good yeast will bring this mixture to a ferment.

A less laborious method of accomplishing the same result is that at one time used in English distilleries. In this a double bottom tub is used, something like that shown in Fig. 41, the upper bottom of which is perforated, and raised above the solid lower bottom. A draw-off cock opens out from the space between the two bottoms.

Assuming that the tub is of 220 gallons capacity, then from 2 to 20 lbs. of chaff are spread over the perforated bottom and pulp from 800 lbs. of raw potatoes placed on that. This is thoroughly drained for half an hour, through the draw-off cock. The pulp is then stirred while from 90 to 100 gallons of boiling water are added gradually. The mass then thickens into a paste. The paste is mashed with about 65 lbs. of well steeped malt, and the liquid left to subside for three or four hours. It is then drained off through the perforated bottom into a fermenting back or tub. For this amount of material the back should be of about 300 gallons capacity.

The leavings left in the preparatory tub still contain considerable starch, and after they are well drained they should be mixed with from 50 to 55 gallons of boiling water. The mixture is then agitated and drained off into the fermenting back. The sediment left is again sprinkled with water, this time cold, which is drained off into the back. This completely exhausts the husks left on the upper bottom. By this process 200 lbs. of potatoes should produce something over 12½ gallons of spirit.

The objection to the last method described is that the spirit so obtained is unpleasant to taste and smell, but this would probably not be an objection for industrial uses.

The only means of obtaining alcohol of good quality from potatoes is to extract the starch separately and then convert it into sugar. This saccharification of the starch may be accomplished by sulphuric acid or by the action of diastase.

By the first of these methods the potatoes are disintegrated in such an apparatus as the Bohn steamer described on page 118. A mixture is made of one-third potatoes, two-thirds water, and one-tenth part of sulphuric acid. The mixture is steamed for six or eight hours under pressure. The mash is then cooled and the acid neutralized by milk of lime. It is then fermented.

By the second and preferable method, dry or wet potato starch is used, which is malted, and the saccharine solution fermented with yeast. The proportions and method for a vat of say 800 gallons capacity are as follows:

Two hundred and sixty-five gallons of water are mixed with 1100 lbs. of dry or 1650 lbs. of moist starch. This mixture is well agitated, and 450 gallons of boiling water run in, together with 165 lbs. of malt. The whole is then stirred energetically and left to saccharify for three or four hours. The saccharine solution thus formed must be brought to 6° or 7° Baume, at a temperature of from 71° to 75° F. To this is then added $1\frac{1}{100}$ lbs. of dry yeast for every 220 gallons of " must." Fermentation is soon established and usually occupies about 36 hours. After remaining at rest for 24 hours the " must " is distilled. From each 220 lbs. of starch there should be a yield of about nine gallons of alcohol, at 90 per cent.

The fermentation of the potato mash is carried on as described in Chapter II. For the preparation of malt see Chapter VI.

CHAPTER VIII.

Alcohol from Grain—Corn, Wheat, Rice, and Other Cereals.

The different cereals constitute a very important source of alcohol in all countries, particularly of course for use in the manufacture of whiskey and gin.

All cereals contain an abundance of starchy substance which under the influence of diastase,—that is, malt,—is converted into fermentible sugar. The quantity of sugar and hence the yield of alcohol differs widely. The following table shows the results obtainable by good workmanship.

220 lbs.	Wheat	gives	7.0	gallons	pure	alcohol
" "	Rye	"	6.16	"	"	"
" "	Barley	"	5.5	"	"	"
" "	Oats	"	4.8	"	"	"
" "	Buckwheat	"	5.5	"	"	"
" "	Corn (Indian)	"	5.5	"	"	"
" "	Rice	"	7.7	"	"	

In addition to these there are other raw materials containing starch which are sometimes used, as millet (55 per cent starch), chestnuts (28 per cent.), and horse chestnuts (40 per cent.). The last is very difficult to work however.

Rice, wheat, rye, barley and corn are more frequently employed than other grains. Wheat gives a malt which is as rich in diastase as barley. Barley and buckwheat are added to these in some proportions. Oats, owing to their high price, are rarely used. Rice, of all the grain is the most productive to the distillers, but on account of its value as a food is not much used for the production of alcohol, unless damaged. Corn is the cereal most largely used for the production of industrial alcohol.

Great care should be exercised in making choice of grain for fermentation where the best results are desired. Wheat should be farinaceous, heavy and dry. Barley should be free from chaff, quite fresh and in large uniform grains of a bright color (see Malting, Chapter VI).

Rice should be dull white in color, slightly transparent, without odor, and of a fresh, farinaceous taste.

The flour or farinaceous part of grain is composed of starch, gluten, albumen, mucilage, and some sugar. The following table gives the proportions of these substances in the commonest grains.

Under certain conditions the albumen or gluten in the grain has the power of converting starch into saccharine matter. This is better effected by an acid such as sulphuric acid, or by a diastase. This latter substance is a principle developed during the germination of all cereals but especially of barley. It has the property of reacting upon starchy matters, converting them first into a gummy substance called dextrine, and then into glucose or grape sugar, see Chapter II.

The action of diastase upon starch or flour made into a paste is remarkable, 50 grains of diastase being sufficient to convert 220 lbs. (100 kilogrammes) of starch into glucose. The rapidity of

TABLE IV.

PROPORTIONS OF STARCH, GLUTEN, ETC., IN PRINCIPAL GRAINS.

Grains.	Starch.	Gluten and other Azotized Substances.	Dextrine, Glucose, and similar Substances.	Fatty Matter.	Cellulose.	Inorganic Salts. (Silica, Phosphates, &c.)
Wheat (average of five varieties)......	65.99	18.03	7.63	2.16	3.50	2.69
Rye....................	65.65	13.50	12.00	2.15	4.10	2.60
Barley	65.43	13.96	10.00	2.76	4.75	3.10
Oats	60.59	14.39	9.25	5.50	7.06	3.25
Indian Corn...........	67.55	12.50	4.00	8.80	5.90	1.25
Rice....................	89.15	7.05	1.00	0.80	1.10	0.90

this change depends on the quantity of water employed, and the degree of heat adopted in the operation.

Inasmuch as barley germinates very readily, and develops a larger proportion of diastase than any other grain, except wheat, it is generally used as a producer of diastase. Barley germinated according to proper methods is called malt. and its preparation is fully described in Chapter VI.

There are many methods of preparing grain for fermentation, but all use at least two of the following operations:—grinding, gelatinizing, steeping, or steaming, mashing saccharifying.

Grinding. Where cookers or the Henze steamers are not used every form of grain should be crushed or ground into a coarse flour. This is in order that the starchy interior may be easily acted on by the diastase. If the grain is not to be mixed with malt later it must be ground more finely so that it may be thoroughly penetrated by the water. The grains should not be ground except as required, as ground grain is liable to heating and consequent loss of fermentability, and is also liable to become musty, in which condition it loses much of its fermentability.

Steeping. This operation is best carried on in vats or tanks of iron or cement, for the reason that wood absorbs impurities, which are communicated to the grain, thus lessening its germinative power. Wooden vats should be thoroughly scrubbed after use, and be kept continually whitewashed. The steeping tub should hold about two-thirds more than the amount of ground grain to be steeped.

Steeping is affected by pouring on to the crushed grain hot and cold water in such quantity that after 10 minutes or so of brewing the mixture will have a temperature of 75° to 95° F.

This warmth makes the water more penetrating. The water should not be poured in all at once, but a little at a time, until the grain is covered to a depth of three or four inches. Care should be taken not to let the temperature get too high, not above 95° F., as a temperature above that point kills the germinating power.

The mixture of crushed grain and water is now stirred for 10 minutes and then left to subside for half an hour. It is then stirred again and the mixture left to steep for 30 or 40 hours, depending on the temperature of the atmosphere, the dryness of the grain, and the character of the water. In very warm weather the water should be changed every few hours by running it off through a hole in the bottom of the tub and running in fresh at the top. This prevents fermentation setting in prematurely.

When the grain swells, and yields readily between the fingers it has been sufficiently steeped, and the water is run off. This is an old method of gelatinizing grain, but a better is by the use of cookers or high pressure steamers as described for potatoes.

Mashing. This consists in mixing the coarse flour with malt and then by means of certain operations and mechanisms bringing it to a condition most favorable to fermentation through the action of yeast. The mixing of the raw flour with barley or other malt effects the conversion of the starch of the grain into maltose. The yeast afterwards converts this maltose into sugar.

Saccharifying. To effect the action of the diastase of the malt on the grain, in the old methods, boiling water must be poured into the vat until the temperature of the mass reaches about 140° to 168° F., the whole being well stirred meanwhile; when this temperature has been reached, the vat is again covered and left to stand for four hours, during which time the temperature should, if possible, be maintained at 140° F., and on no account suffered to fall below 122° F., in order to avoid the inevitable loss of alcohol consequent upon the acidity always produced by so low a temperature. In cold weather the heat should of course be considerably greater than in hot. It should be also remarked that the greater the quantity of water employed, the more complete will be the saccharification, and the shorter the time occupied by the process.

Having undergone all the above processes, the wash is next drawn from the mash tub into a cistern, and from this it is pumped into the coolers. When the wash has acquired the correct temperature ,viz., from 68° to 78° F., according to the bulk operated upon, it is run down again into the fermenting vats situated on the floor beneath. Ten to twelve pints of liquid or 5½ to 6½ lbs. of dry brewer's yeast are then added for every 220 lbs. of grain; the vat is securely covered, and the contents are left to ferment. The process is complete at the end of four or five days, and if conducted under favorable conditions there should be a yield of about 11 gallons of pure alcohol to every 220 lbs. of grain employed.

There are a number of different methods of mashing, having each its advantages, and applicable to particular varieties of grain.

We will first consider the mashing of the steeped grain in general by one of the older and simpler processes.

The grain to be mashed, which has been ground and steeped as before described, is mixed with malt in the proportion of four to one, or even eight to one. In addition, three or four pounds of chaff to every hundred or so pounds of steeped grain should be used.

Mash. Water is then run into the mash tub in the proportion of about 600 gallons to each 60 bushels of grain. Its temperature should be between 120° and 150° F. During the entrance of water, the mass is well stirred so as to cause the whole of the grain to be thoroughly soaked and to prevent the formation of lumps. It is best to add the grain to the water gradually and to stir thoroughly.

To this mass about 400 gallons of boiling water is gradually added to keep the temperature at about 145° F. During the addition of the boiling water the mash should be continually stirred so that the action of the water shall be uniform. This operation should last about two and one half hours. The vat should be then covered and left to stand from three-quarters to one hour for saccharification.

Another method of saccharifying is to turn boiling water gradually into the mash tank until the mixture has acquired a temperature of from 140° to 180° F. The mass is thoroughly stirred, and the tub is covered and left to subside for from two to four hours, during which time the temperature should not be allowed to fall below 120° F. A small tub needs more heat than a larger tub, and more heat is required in winter than in summer.

A convenient method of regulating the temperature of the mash tank, would be by a coil of pipes on the bottom. This would be connected by a two-way cock to a steam boiler and to a source of cold water. Heat should never be carried over 180° F., and the best temperature is from 145° to 165° F.

The greatest effect of the diastase of the malt upon the gelatinized starch is at 131° F. For ungelatinized starch this is not great enough, hence the greater part of the mashing is carried on at the lower temperature and only towards the end should the temperature be raised to the maximum 150° F.

Every distiller uses his own judgment as to the amount of the mashing water used, its temperature, the length of time during which the mash rests, and the length of time for saccharification.

Saccharification may be recognized by the following signs: The mash loses its first white mealy look, and changes to dark brown. It also becomes thin and easily stirred. The taste is sweet and its odor is like that of fresh bread.

Corn and other grain may be mashed conveniently in such an apparatus as that described on page 1C, as used for potatoes the steam being introduced under pressure.

The water is first placed in the steamer. Steam is introduced into the water and it is brought to a boil The corn is then introduced gradually, the steam pressure increased to its maximum, and the mass blown out as described in Chapter VII. Hellefreund's apparatus (see page 118) may also be used with ground corn.

The corn or grain not previously crushed or ground is introduced into a steamer in the proportion of 200 lbs. of corn to 40 gallons of water. The steamer should have about 100 gallons of steam space for this amount.

The mashes described above are thick, more or less troublesome to distil, and only simple stills can be used. By the following method a clear saccharine fluid or wort can be obtained.

A mash vat is used having a double bottom. The upper bottom is perforated and between the two bottoms is a draw-off pipe and a pipe for the inlet of water.

Upon the upper perforated bottom is first placed a layer of between two and three pounds of chaff. Upon this is turned in a mixture of 400 lbs. corn and malt in the proportions of $\frac{1}{5}$ malt to $\frac{4}{5}$ grain. Eighty-seven gallons of water at a temperature of from 85° to 105° F. is then let in to the bottom, while the mixture is thoroughly agitated for 10 minutes. It is then left to subside for half an hour.

After this steeping process, the mass is again agitated while 175 gallons of water at 190° F. are let into the tub while the mass is continually and thoroughly stirred by mechanical stirrers. Brewing lasts for half an hour, and the liquid is then left to stand for seven hours.

At the end of this period the grain is covered by clear liquid which is drained off through the draw-off cock into the fermenting back.

To the contents left in the steeping tank 135 gallons of boiling water are added as before and the liquid therefrom drawn into the fermenting back.

It usually requires three infusions to extract the whole of the saccharine and fermentiscible matters contained in the grain. In some places, it is customary to boil down the liquors from the three mashings until they have acquired a specific gravity of about 1.05, the liquor from a fourth mashing being used to bring the whole to the correct degree for fermentation, the liquors from the third and fourth being boiled down to the same density and then added to the rest. In a large Glasgow distillery, the charge for the mash tubs is 29,120 lbs. of grain together with the proper proportion of malt. Two mashings are employed, about 28,300 gallons of water being required; the first mashing has a temperature of 140° F., and the second that of 176° F. In Dublin the proportion of malt employed is only about one-eighth of the entire charge. One mashing is employed, and the temperature of the water is kept at about 143° F. The subsequent mashings are kept for the next day's brewing.

By this process the grain is entirely deprived of all fermentible substances which have been carried away in a state of liquid sugar.

The whole operation of preparing and saccharifying grain is to-day carried on in steamers, such as described on page 11, and cooking apparatus such as shown in Fig. 1, or in the Henze high pressure steamers and preparatory mash vats described in Chapter II.

In steaming grain without pressure, the finely crushed grain is poured slowly into a vat previously nearly filled with water at a temperature of about 140 degrees F. A little less than half a gallon of water is used for each pound of grain. Care must be taken to stir the mass constantly to prevent lumping. When all the corn is mixed in, steam is allowed to enter and the temperature raised to about 200 degrees F. It should be left at this temperature for an hour, or an hour and a half, when the temperature is reduced to 140° F. when about 10 per cent. of crushed malt is added and the temperature reduced to 68° F. by means of suitable cooling devices.

When steam cookers are used, the cylindrical boiler is first filled to the proper degree with water at a temperature of 140° F. The meal is then let in gradually being constantly stirred the while. The boiler is then closed and steam gradually let in while the mass is stirred until a pressure of 60 pounds and a temperature of 300° F. has been reached. The starch then becomes entirely gelatinized, the pressure is relieved, and the temperature reduced to 212° F. and then rapidly brought to 145° F. The malt is added mixed with cold water, at such a stage before the saccharifying temperature is reached that the cold malt and water will bring it to 145° F. The malt is stirred and mixed with the mash for five or ten minutes and the mixed mass let into a drop tub when saccharification is completed. It is then cooled as described.

When the Henze steamers are used the grain may be treated in either the whole grain or crushed, as the high pressure to which it is subjected and the " blowing out " act to entirely disintegrate it. In this mode of operation, water is first let into the steamer and brought to a boil by the admission of steam. The grain is then slowly let into the apparatus. The water and grain should fill the steamer about two thirds full. The steamer is left open and steam circulated through the grain and water for about an hour, but without any raising of pressure. This acts to thoroughly cook and soften the grain.

When sufficiently softened the steam escape cock in the upper part of the steamer (see Fig. 2) is regulated to allow a partial flow of steam through it and a greater flow of steam is admitted though the lower inlet. This keeps the grain in constant ebullition under a pressure of 30 lbs. or so. After another period of an hour the pressure in the steamer is raised to 60 lbs. at which point it is kept for half an hour, when the maximum steam pressure is applied, and the greater portion of the disintegrated mass blown out into a preparatory mass tub, into which malt has heen placed mixed with water. The blowing out should be so performed that the temperature in the mass in the tubs shall not exceed 130° F. The mass is stirred and cooled and then the remainder of the mass in the steamer admitted to the tub which should bring the temperature of the mass up to 145° F. It is kept at this temperature for a period varying from half an hour to one and one-half hours and is then cooled to the proper fermenting temperature.

Another method of softening corn so that its starch is easily acted upon by the diastase of the malt is to steep it in a sulphurous acid solution at a temperature of about 120° F. for from fifteen to twenty hours. The mass is then diluted to form a semi-liquid pulp and heated to about

190° F. for an hour or two during which the mass is constantly stirred. The malt is then added, the mass is saccharified, cooled and then fermented.

Another method is to place mixed grain and hot water in a cooker of the Bohn variety (Fig. 45). After half an hour of stirring and cooking under ordinary pressure, the steam pressure is raised to 45 lbs. This is kept up for from two to three hours when the grain is reduced to a paste. Concentrated muriatic acid equal to 2½ per cent of the weight of grain is then forced in, under steam pressure. In half an hour the grain will be entirely saccharified and ready for fermenting.

CHAPTER IX.

ALCOHOL FROM BEETS.

Cultivation. The beetroot (*Beta vulgaris*), indigenous to Europe, is cultivated in France, Germany, Belgium, Holland, Scandinavia, Austria, Russia, and to a very small extent in England and New Zealand, and to a very large extent in the United States and Canada. There are many varieties. The most important to the sugar-maker is the white Silesian, sometimes regarded as a distinct species (*B. alba*); it shows very little above ground, and penetrates about 12 in.; it has a white flesh, the two chief forms being distinguished by one having a rose-colored skin and purple-ribbed leaves, the other a white skin and green leaves. Both are frequently grown together, and exhibit no marked difference in sugar-yielding qualities.

Good sugar-beets possess the following broad characteristics: (1) Regular pear-shaped form and smooth skin; long, tapering, carrot-like roots are considered inferior; (2) white and firm flesh, delicate and uniform structure, and clean sugary flavor; thick-skinned roots are spongy and watery; those with large leaves are generally richer; (3) average weight 1½ to 2½ lbs., neither very large nor very small roots being profitable to the sugar-manufacturer; as a rule, beets weighing more than 3½ lbs. are watery, and poor in sugar; and roots weighing less than ¾ lb. are either unripe or too woody, and in either case yield comparatively little sugar; the sp. gr. of the expressed juice, usually 1.06 to 1.07, even reaching 1.078 in English-grown roots, indicating over 14 per cent. of crystallizable sugar, is the best proof of quality; juice poor in sugar has a density below 1.060; (4) in well-cultivated soil, the roots grow entirely in the ground, and throw up leaves of moderate size.

Composition of the Roots. Internally the root is built up of small cells, each filled with a juice consisting of a watery solution of many bodies besides sugar. These include several crystallized salts (mostly of which are present in minute traces only), such as the phosphates, oxalates, malates, and chlorides of potassium, sodium, and calcium, the salts of potash being by far the most important; and several colloid bodies (albuminous [nitrogenous] and pectinous compounds); as well as a substance which rapidly blackens on exposure to the air. The greater part of the sugar in ripe beets is crystallizable, and, when perfectly pure, is identical in composition and properties with crystallized cane-sugar; but it is more difficult to refine this sugar so as to free it from the potash salts, and commercial samples have not nearly so great sweetening power as ordinary cane-sugar.

Beets contain no uncrystallizable sugar; the molasses produced in beet-sugar manufactories is the result of changes which cannot be entirely avoided in extracting the crystallizable sugar.

Soil. The best soil for beets contains a fair proportion of organic matter, is neither too stiff nor too light, and crumbles down into a nice friable loam; it must be capable of being cultivated to a depth of at least 16 in. The subsoil should be thoroughly well drained, and rendered friable by autumn-cultivation and free admission of air. A deep friable turnip-loam, containing fair proportions of clay and lime, appears to be the most eligible land for sugar-beets. Lime is a very desirable element. Well-worked clay-soils, especially calcareous clays, are well adapted, if properly drained and of sufficient depth. Peaty soils and moorlands are quite unsuitable, as well as lands which are too dry, like the thin gravelly soils resting on siliceous gravel sub-soils, or too wet and cold, like many of the thin soils above impervious chalk marl.

Speaking generally, the best soils for sugar-beet are precisely those on which other root-crops can be grown to perfection, that is, land which is neither too heavy nor too light, which has a good depth, is readily penetrated by the roots, and naturally contains lime, potash, clay, and sand, as well as organic matter, is such proportions as in good friable clay-loams. An analysis of the soil should be made previous to planting it with the sugar-beet, as the salts presented in solution in the soil will pass into the juice, and greatly interfere with the processes of sugar manufacture. Certain soils may be at once indicated as unsuitable; they are clover-land, recent sheep pastures, forest-land grubbed during the preceding 15 years, the neighborhood of salt works, volcanic and saline soils of all kinds. The beet requires a certain supply of potash and soda salts in the soil, but if these are present in excess, as in recent forest-land, the juice does not work well, nor give its proper yield of sugar.

Manures. Sugar-beets should be grown with as little farmyard manure as possible; when dung has to be used, as in the case of very poor soils, it should be applied in autumn, or as early as possible during the winter months. The effect of heavy dressings of animal nitrogenous matters or ammoniacal salts, is to produce abundance of leaves, and big watery roots; the latter are comparatively poor in sugar, and contain potash salts derived from the animal matters, which greatly interfere with the extraction of sugar in a crystallized state. Common salt, and saline manures in general, though useful in moderate doses (224 lbs. to 336 lbs. per acre on light soils), should be avoided on the majority of soils, for sugar-beets grown on soils highly manured with common salt produce juice largely impregnated with salt, which is dreaded by the manufacturer even more than albuminous impurities, and nearly as much as excess of potash salts.

If the land is in good condition, containing sufficient available nitrogen to meet the requirements of the crop, neither guano nor sulphate of ammonia should be used. They largely increase the weight of the produce per acre; but heavy crops are generally poor in sugar, and furnish a juice that presents much difficulty to the manufacturer. If the land is very poor, and if farmyard manure cannot be obtained and be applied in autumn, 336 to 448 lbs. of Peruvian guano, or 224 lbs. of sulphate of ammonia, mixed with 224 lbs. of superphosphate of lime, per acre, may be sown broadcast in autumn, and 224 lbs. more of superphosphate may be drilled in with the seed in spring. Superphosphate of lime and bones are excellent for sugar-beets, and never injure the quality of the crop, like the indiscriminate use of ammoniacal manures. On light soils, in which potash is often deficient, the judicious use of potash salts has been found serviceable, but only in conjunction with superphosphate and phosphatic guanos.

Sowing. The best time for sowing beetroot is the beginning or middle of April. If sown too early, the young plants may be partially injured by frost; if later than the first week in May, the crop may require to be taken up in autumn, before it has had time to get ripe. About 10 to 12 lbs. of seed is required per arce. As regards the width between the plants, generally speaking, the distance between the rows and from plant to plant should not be less than 12 nor greater than 18 in. Should the young plants be caught by a night's frost, and suffer ever so little, it is best to plough them up at once and re-sow, for they are certain to run to seed, and are then practically useless for the manufacture of sugar. Sugar-beets require to be frequently horse- and hand-hoed. As long as the young plants are not injured, the application of the hoe from time to time is attended with great benefit to the crop. It is advisable to gather up the soil round each plant, in order that the head may be completely covered with soil. Champonnois' researches point to the advantage of planting in ridges, by which the supply of air to the roots is greatly facilitated.

The conditions best calculated to ensure the roots possessing the characters most desirable from a sugar-maker's point of view are chiefly as follows: (1) Not to sow on freshly-manured land; it is eminently preferable not to manure for the beet crop, but to manure heavily for wheat in the preceding year; (2) not to employ forcing manures, nor to apply manure during growth; (3) to use seed from a variety rich in sugar; (4) to sow early, in lines 16 in. apart, at most, the plants being 10 to 11 in. from each other; there will then be 38,000 beets on an acre, weighing 21 to 28 ounces each, or 52,800 to 70,400 lbs. per acre; (5) to weed the fields as soon as the plants are above ground, to thin out as early as possible, and to weed and hoe often, till the soil is covered with the leaves of the plants; (6) never to remove the leaves during growth; (7) finally, not to take up the roots, if it can be avoided, before they are ripe, the period of which will depend upon the season.

Good seed may be raised by the following means: The best roots, which show least above ground, are taken up, replanted in good soil, and allowed to run to seed. This seed is already good; but it may be further improved by sowing it in a well-prepared plot possessing all the most favorable conditions; the resulting plants are sorted, set out in autumn, put into a cellar, and in the spring, before transplanting, those of the greatest density, and which will give seeds of the best quality, are separated. These are transplanted at 20 in. between the rows and 13 in. between the feet, which are covered with about 1½ in. of earth. Finally they are watered with water containing molasses and superphosphate of lime, as recommended by Corenwinder.

Harvesting. Sugar-beets must be taken up before frost sets in. When the leaves begin to turn yellow and flabby, they have arrived at maturity, and the crop should be watched, that it may not get over-ripe. If the autumn is cold and dry, the crop may be safely left in the ground for seven to ten days longer than is needful, but should the autumn be mild and wet, if the roots are left in the soil, they are apt to throw up fresh leaves, and nothing does so much injury. In watching the ripening of the crop, a good plan is to test the sp. gr. of the expressed juice. A root or two may be taken up at intervals, and reduced to pulp on an ordinary hand-grater, the juice obtained by pressing the pulp through calico, and the density observed by a hydrometer. As long as the gravity of the juice continues to increase, the crop should be left in the land. Good sugar-yielding juice has a sp. gr. of about 1.065, rising to about 1.070. Immature roots, cut across, rapidly change color on the exposed surface, turning red, then brown, and finally almost black. If newly-cut slices turn color on exposure, the ripening is not complete; but if they remain some time unaltered, or turn only slightly reddish, they are sufficiently ripe to be taken up. The crop should be harvested in fine, dry weather. In order that the roots may part with as much moisture as possible, they are left exposed to the air on the ground before being stacked, but not for longer than a few days, and they need to be guarded against direct sunlight. Perhaps the best plan is to cover them loosely with their tops in the field for a couple of days, then trim them, and at once stack them.

Storing. For storing roots, especial care should be taken to prevent their germinating and throwing out fresh tops, which is best done by selecting a dry place for the storage ground. They may be piled in pyramidal stacks, about six feet broad at base, and seven feet high. At first, the stacks should be thinly covered with earth, that the moisture may readily evaporate; subsequently, when frosty weather sets in, another layer of earth, not exceeding one foot in thickness, may be added. This is essentially the method generally adopted for storing potatoes and beets.

In continental Europe and Canada, extra pre-

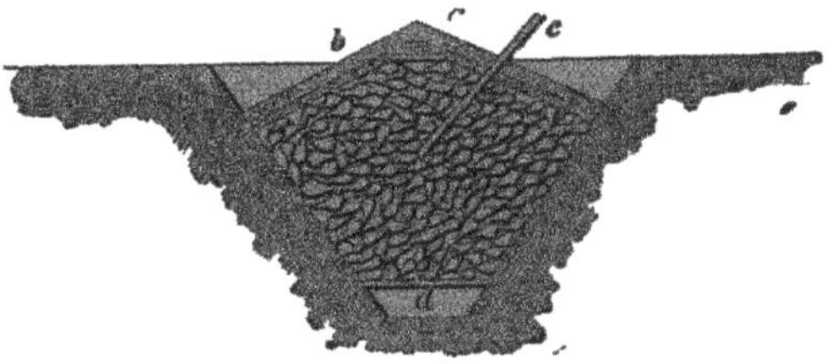

FIG. 46.—Stack for Storing Beets.

caution is necessitated by the rigorous climate. In S. Russia, the plan shown in Fig. 46 is sometimes used. The beets are disposed completely below the surface of the soil, in a trench dug with sharply sloping sides. At about 15 in. from the bottom, is an openwork floor of reeds, on which the beets are piled to within a few inches of the level of the exterior soil. On the top, and following the apex of the heap, is laid a triangular ridge-piece *a*, for the purpose of facilitating evaporation. The whole is covered with a layer *b* of straw and tine earth, the thickness of which is varied accroding to the indications of the thermometer *c* placed in the center of the mass. Between the floor of the trench and the openwork floor is a space *d*, communicating with two vertical channels leading to the outer air, thus providing ventilation. The outlets of the channels can be opened and closed at will. The Russians also often employ regular cellars, as shown in Fig. 47. The structure con-

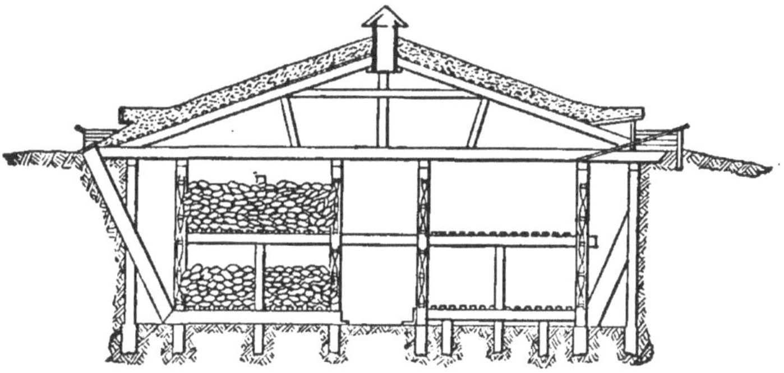

FIG. 47.—Storage Cellar for Beets.

sists of two stories, covered with a bed of earth, each furnished with a floor of hurdles or open planking, on which the beets are piled to the depth of about one yard. Lateral passages facilitate ventilation, and openings in the roof permit the heated air to escape. The cost of erecting these cellars is heavy, but there is great saving of labor in storing the beets, as it suffices to simply pile them up on the floors. Moreover, the arrangement permits the examination of the contents beyond the indications of a thermometer; and enables any portion to be removed, even during snowy weather.

Alcohol from Beets. Beets contain 85 per cent. of water, and about 10 per cent. of cane sugar, the remainder being woody fibre and albumen; cane sugar not being in itself fermentible,—as is grape sugar,—it has to be converted into "inverted sugar" by a ferment as yeast. Either the sugar beets may be mashed or the molasses which remains from the manufacture of beet sugar (as described in Chapter X). The conversion of the sugar into alcohol is effected in several different ways, of which the following are the principal:

By rasping the roots and submitting them to pressure, and fermenting the expressed juice.

By maceration with water and heat.

By direct distillation of the roots.

The first two methods are the best as by them the woody fibre of the plant which is non-fermentible is separated from the fermentible juice. In both the first and second processes the beets must first be entirely cleaned of adhering dirt, trash and clods of earth, and then rasped, pulped or sliced by certain machinery.

Cleaning. Care must be taken in this operation that the beets shall be freed from small stones and adhering hard lumps of earth which would otherwise get into the rasping machinery to the damage and stoppage of the mechanism.

There are many forms of cleaners but all are alike in this,—that the beets shall be subjected to the action of water while traveling through or over a perforated casing. The simplest machine, and one easily constructed by any carpenter, comprises an elongated cylinder formed of lathes or strips spaced apart such distance as will allow dirt and stones to pass between them. This is mounted on a central shaft and revolves in a tank of water. It should be slightly inclined so that the potatoes or beets to be washed may feed downward from the open upper end-disk or wheel, to the lower end where they are thrown out. At the upper end is a hopper and at the lower, the end disk has inwardly projecting lips, which as the cylinder revolves lifts the beets up and tumbles them out on to an incline which carries them to the rasping machine.

Another form of machine comprises a perforated cylinder of sheet iron, revolving in a tank of water. A better form of cleaner than either of those consists of an inclined trough in which a spiral feeding screw of sheet iron rotates. The beets are fed into the trough at its lower end and are carried upward, slowly, by the feeding screw. Above the trough is a water pipe having a number of outlets by which water may fall on to the beets and into the trough. The water rushing down the inclined trough carries with it all dirt and stones, and by the time the beets have reached the upper end they are entirely cleaned and ready for slicing or rasping.

For pressing out the juice, the beets are mashed into a pulp, while for diffusion the beets are sliced.

Rasping. Fig. 48 shows one form of rasping machine. On a suitable supporting frame is mounted a cylinder *a* having a diameter of about 24 inches. The cylinder is formed of alternate saw blades and wooden washers holding them a slight distance apart. The saws or teeth are so set on the cylinder as not to slice the beets but to shred them up into a fine pulp. The cylinder rotates at a speed of 800 to 1000 revolutions a

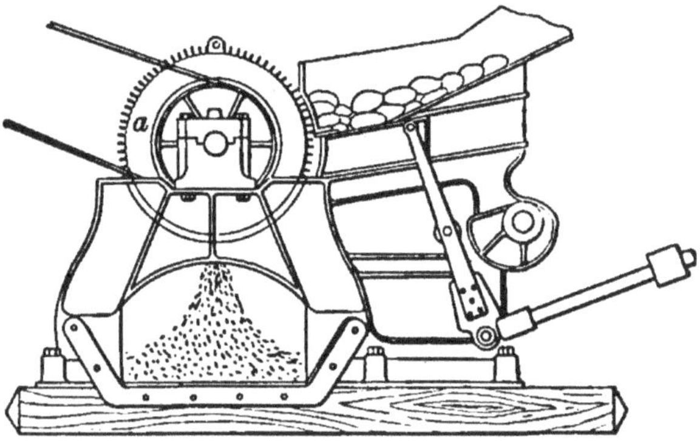

FIG. 48.—Beet and Potato Rasp.

minute in front of an inclined table, having a jigger whereby the beets are fed downward against the toothed cylinder. The teeth carry the pulp downward and it falls into a receptacle beneath.

It is best to add to this pulp a small portion of sulphuric acid, say two-tenths of one per cent. This prevents by-fermentations.

Pressing. The pulp obtained from the raspers has now to be expressed. This is either done by platen presses or by roller presses. With platen presses the first pressing may be done by screws, but the final pressing should be accomplished by hydraulic presses.

For the hydraulic press, the pulp is placed in woolen sacks, containing 10 to 12 lbs., superposed in the press with their mouths doubled under, and separated by iron plates; about 25 are collected, and the pile is put into a screw-press, called a " preparatory " press, which extracts about 45 to 50 per cent. of the juice. These pressed sacks are piled anew on the movable plate of a powerful hydraulic press, which takes 50 at a charge. Each preparatory press can supply four hydraulic presses, which are ranged around it, so that of the four presses, there will be one charging, one commencing to press, one in full pressure, and one discharging, at the same moment. Motion is communicated to the four hydraulic presses by four pumps mounted on the same bed, and tended by the same workman who directs the pressing. An improvement upon the general form of hydraulic press is that devised by Lalouette, which enables two workmen and one boy to work five presses. These presses turn out about 34,200 lbs. per 24 hours in the first pressing, and 68,400 lbs. in the second. Hydraulic presses are rapidly falling into disuse in the beet-sugar industry, by reason of the superior merits of continuous presses, and the extended adoption of the diffusion system.

Continuous presses for beet were suggested by the roller-mills used in the cane-sugar industry. But the conditions in the two cases are widely different; the begass of the cane is solid, and readily parts from the juice; whereas the pulp and juice of the beet have a strong tendency to combine, and the roller-surface must therefore be permeable only by the juice. In Poizot et Druelle's press, the pulp passes between two cylinders, carried by endless cloths. The object is to unite the best features of the hydraulic press. To this end, a first gentle pressing is produced against the first cylinder by the elasticity of the principal cloth on which it is borne. Then, encountering a series of four little rollers, performing the functions of the preparatory press, it is next seized between the second and first cylinders, and deprived of the maximum quantity of juice.

Dujardin's roll press is shown in Fig. 49, which is a vertical section of the machine, the side plate being removed. The pulp is forced upward through a pipe *C* under high pressure. This has a regulating slide valve *D*. The rolls *B B* revolve towards and nearly in contact with each other, and they are perforated so that the expressed juice may run off through the rolls. These perforations are conical in form with the apex of the cone outward. The cylinders are also covered with a webbing of cloth or horse hair. Below the rolls is block *C'*, which with the outer walls of the chamber, form diverging passages which extend upward, as shown, on either side of the rolls and then downward

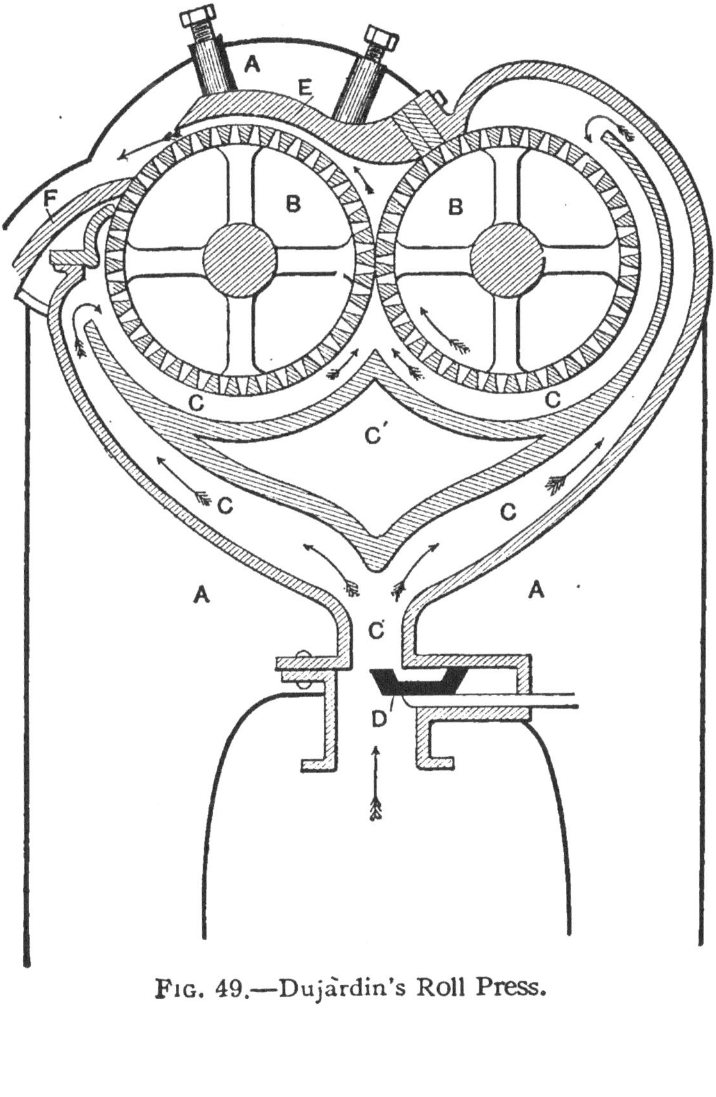

FIG. 49.—Dujardin's Roll Press.

along the lower faces of the rolls to the point when they contact. The pulp is compressed with great force against and between the rolls, the juice is forced through the perforations and the residue passes upward and outward under the presser bar *E* in the form of a ribbon which is guided away by the trough *F*. The pressure of the bar *E* is regulated by screws and the tighter said bar is pressed against the rolls the greater will be the pressure of the pulp behind the bar and against the rolls, and the greater the juice expressed.

The rolls revolve very slowly only about seven or eight times a minute but the capacity of the machine is very great, it being capable of pressing the pulp of from 85,000 to 175,000 lbs. of beets daily. The residue from the first pressing should be submitted to a further pressing after being macerated with spent wash. This residue may be fed to cattle. The utmost cleanliness is essential to these processes; all the utensils employed should be washed daily with lime-water to counteract acidity.

Extraction by Maceration and Diffusion. The object of this process is to extract from the beets by means of water or spent liquor all the sugar which they contain, without the aid of rasping or pressure. Spirit is thus produced at considerably less expense, although it is not of so high a quality as that yielded by the former process. The operation consists in slicing up the beets in a specially constructed slicing machine, into slices of regular thickness, and then allowing the slices to macerate in a series of vats at stated temperatures. It is essential that the knives by which the roots are cut should be so arranged that the roots are divided into slices having a width of $\frac{4}{10}$ of an inch and a thickness of $\frac{4}{100}$ of an inch, and a variable length; the roots are, of course, well washed before being placed in the hopper of the cutter.

When cut, the beets are covered with boiling water in a macerator of wood or iron for one hour, the water should contain 4.4 of sulphuric acid to every 2200 lbs. of beets. After this, the water is drawn off into a second vat in which are placed more beets, and allowed to macerate again for an hour. This is repeated a third time in another vat, and the juice, which has now acquired a density equal to that obtained by rasping, is run off into the fermenting vat. When the first vat is empty it is immediately refilled with boiling water and fresh beets; the juice from this operation is run into the second vat, when the contents of that one are run into the third. To continue the operation, the beets are completely exhausted by being macerated for an hour with a third charge of boiling water (acidulated as in the former case). The exhausted pulp is removed to make room for fresh slices; and the first vat is then charged with juice which has already passed through the second and third vats. After macerating the fresh beets for one hour, the charge is ready for fermentation. In ordinary weather, the juice should now be at the right heat for this process, viz., about 71.1° or 75.2° F., but in very cold weather it may require some re-heating.

In Fig. 50 is shown a series of vats for the extraction of the sugar from beets such as is termed a " diffusion battery."

The vessels, 1, 2, 3 and 4 are of wood or sheet

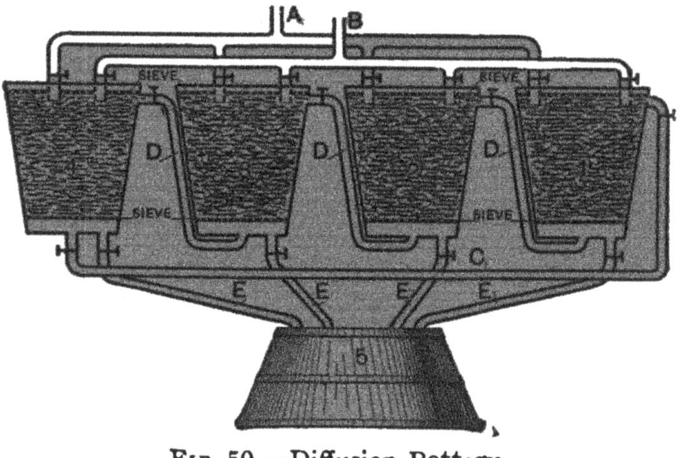

Fɪғ. 50.—Diffusion Battery.

iron. Each vessel has a bottom sieve and a top sieve between which the beet slices are to be placed. From the bottom of each vessel below the sieve a pipe *D* runs to the top of the vessel next in order. From the bottom of the last vessel 4 of the series a pipe *C* runs back to the top of the one first used. Pipes *A* and *B* are connected to each vessel for the admission of water and spent wash respectively. A discharge pipe *E* leads from each vessel to a collecting vat 5.

Maceration and diffusion is accomplished as follows: The sliced beets are placed between the sieves in vessel 1 and water or spent wash at a temperature of 185° F. is let in and the beets allowed to macerate for three-quaiters of an hour, meanwhile tub 2 is charged with sliced beets. The cock or pipe *D* between the vessels is opened when the time, three quarters of an hour, has elapsed; hot water or spent wash is admitted by pipes *A* or *B* to the vessel 1, which forces the sugar solution therein into vessel 2. When the required amount of fluid has been passed into 2 from 1, the inlet of water into 1 is stopped, and the vessel heated to 185° F.

Vessel 3 is charged with beet slices and in three-quarters of an hour vessels 1, 2 and 3 are connected and water or wash admitted into 1, which forces the solution in 1 into 2 and that in 2 into 3 when it is again raised to 185° F.

The same operation is repeated as to vessel 4 and in three-quarters of an hour all the vessels are connected, hot water or spent wash is admitted to 1 and the sugar solution drawn off from 4 into the vat.

The beets in tub 1 having now been exhausted, the fluid in that vessel is drawn off and the exhausted beets thrown away. 1 is now recharged with beets and the pipe between it and 4 opened. The former operation is repeated except that now vessel 4 becomes 1, and 1 becomes 4. These successive chargings and dischargings are continued; vessel 3 becomes 1 in its turn and so on.

Fermentation. Before fermentation the juice procured as has been described is brought to about 82° F.; at this temperature it is run off into the fermenting vats. Here it is necessary, as before noted, to add to the juice a small quantity of concentrated sulphuric acid, for the purpose of neutralizing the alkaline salts which it contains, and of rendering it slightly acid in order to hasten the process; this quantity must not exceed 5½ lbs. to every 1220 gallons of juice, or the establishment of fermentation would be hindered instead of promoted. The addition of this acid tends also to prevent the viscous fermentation to which the juice obtained by rasping and pressure is so liable. Although the beet contains albumen, which is in itself a ferment, it is necessary, in order to develop the process, to have recourse to artificial means. A small quantity of brewer's yeast—about 1¾ oz. per 22 gallons of juice—is sufficient for this; the yeast must previously be mixed with a little water. An external temperature of about 68° to 78° F. must be carefully maintained. Fermentation lasts for from four to five hours.

The fermentation of acidulated beet-juice sets in speedily. The chief obstacle to the process is the mass of thick scum which forms upon the surface of the liquor. This difficulty is sometimes obviated by using several vats and mixing the juice, while in full fermentation, with a fresh quantity. Thus, when three vats are employed, one is set to ferment; at the end of four or six hours, half its contents are run into the second vat and here mixed with fresh juice. The process is arrested, but soon starts again in both vats simultaneously; the first is now allowed to ferment completely, which is effected with much less difficulty than would have been the case had the vat not been divided. Meanwhile the second vat, as soon as the action is at its height, is divided in the same manner, one-half its contents being run into the third. When this method is employed, it is necessary to add a little yeast from time to time when the action becomes sluggish.

Direct Distillation of the Roots. This process, commonly called "Leplay's method," consists in fermenting the sugar in the slices themselves. The operation is conducted in huge vats, holding as large a quantity of matter as possible, in order that the fermentation may be established more easily. They usually contain about 750 gallons, and a single charge consists of 2200 lbs. of the sliced roots. The slices are placed in porous bags in the vats, containing already about 440 gallons of water acidulated with a little sulphuric acid; and they are kept submerged by means of a perforated cover, which permits the passage of the liquor and of the carbonic acid evolved; the temperature of the mixture should be maintained at about 77° or 80° F. A little yeast is added, and fermentation speedily sets in; it is complete in about 24 hours or more, when the bags are taken out and replaced by fresh ones; fermentation declares itself again almost immediately, and without any addition of yeast. New bags may, indeed, be placed in the same liquor for three or four successive fermentations without adding further yeast or juice.

The slices of beets charged with alcohol are now placed in a distilling apparatus of a very simple nature. It consists of a cylindrical column of wood or iron, fitted with a tight cover, which is connected with a coil or worm, kept cool in a vessel of cold water. Inside this column are arranged a row of perforated diaphragms or partitions. The space between the lowest one and the bottom of the cylinder is kept empty to receive the condensed water formed by the steam, which is blown into the bottom of the cylinder in order to heat the contents. Vapors of alcohol are thus disengaged from the undermost slices, and these vapors as they rise through the cylinder vaporize the remaining alcohol, and finally pass out of the top at a considerable strength and are condensed in the worm. When all the contents of the still have been completely exhausted of spirit, the remainder consists of a cooked pulp, which contains all the nutritive constituents of the beet except the sugar.

CHAPTER X.

Alcohol from Molasses and Sugar Cane.

Another common source of alcohol is molasses. Molasses is the uncrystallizable syrup which constitutes the residiuum of the manufacture and refining of cane and beet sugar. It is a dense, viscous liquid, varying in color from light yellow to almost black, according to the source from which it is obtained; it tests usually about 40° by Baume's hydrometer. The molasses employed as a source of alcohol must be carefully chosen; the lightest in color is the best, containing most uncrystallized sugar. The manufacture is extensively carried on in France, where the molasses from the beet sugar refineries is chiefly used on account of its low price, that obtained from the cane sugar factories being considerably dearer. The latter is, however, much to be preferred to the former variety as it contains more sugar. Molasses from the beet sugar refineries yields a larger quantity and better quality of spirit than that which comes from the factories. Molasses contains about 50 per cent. of saccharine matter, 24 per cent. of other organic matter, and about 10 per cent. of inorganic salts, chiefly of potash. It is thus a substance rich in matters favorable to fermentation. When the density of molasses has been lowered by dilution with water, fermentation sets in rapidly, more especially if it has been previously rendered acid. As, however, molasses from beet generally exhibits an alkaline reaction, it is found necessary to acidify it after dilution; for this purpose sulphuric acid is employed, in the proportion of about 4½ lbs. of the concentrated acid to 22 gallons of molasses, previously diluted with eight or ten volumes of water, Three processes are thus employed in obtaining alcohol from molasses; dilution, acidification, and fermentation. The latter is hastened by the addition of a natural ferment, such as brewer's yeast. It begins in about eight or ten hours, and lasts upwards of 60.

About three gallons of Alcohol may be obtained from one hundred pounds of molasses.

Beet Sugar Molasses. The first step in the process of rendering the molasses fermentable is to mix the molasses with water, to a certain dilution, in the proportion of two parts of water to one of molasses. This may be done by hand, but preferably it is performed in a vat provided with stirring or agitating mechanism, such as will effectually mix the water with the viscid syrup, and whereby also the wash may be thoroughly agitated and aerated.

There are numerous forms of mixing vats, all working however, on the principle shown in Fig. 51. In this, the vat *A* is provided with a central shaft *C* carrying radial mixing blades *E*. This shaft is driven by bevel gears *D*, *F*. As the rotation of these blades would merely tend to create a rotary current of molasses and water, and not to mix them, some means should be used for impeding and breaking up this current. To that end the cover is provided with downwardly projecting rods *I* which create counter currents, and thoroughly intermingle the two liquids. Another and even better form of mixer consists of a tank into the

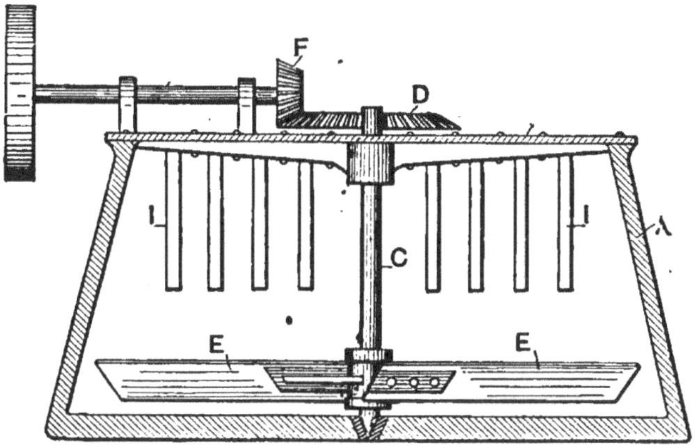

FIG. 51.—Mixing Vat.

lower portion of which enters a perforated pipe of relatively large diameter. This is provided at the end with an air entrance and a steam injector. The injected steam draws in air and the steam and air are forced under pressure into the vat, thus diluting the contained molasses, agitating it and thoroughly aerating it.

The molasses as it comes from the sugar house may

contain anywhere from 30 to 45 per cent of sugar, and this should be diluted with water to a concentration of 16 to 18 per cent of sugar.

The density of the wash after " setting up " is 1.060. It is to be noted that though with improved apparatus a wash as concentrated at 12° or 15° Baume may be worked; yet where simple apparatus is used six degrees or eight degrees is better and much more favorable to rapid and complete fermentation.

After setting up, one gallon of strong sulphuric acid and 10 lbs. of sulphate of ammonia are added for each 1000 gallons of wash. This neutralizes the alkaline carbonates in the beet juice which would otherwise retard fermentation, and it assists the yeast to invert the cane sugar as formerly described. The addition of ammonia is in order to give food to the yeast and obtain a vigorous fermentation.

The yeast used for fermenting molasses is prepared either from malt or grain and is used as concentrated as possible, and in the proportion of about 2 per cent.

The " pitching " temperature of a molasses wash varies with the concentration of the wash, being higher for strongly concentrated solutions than for weak ones. When the wash tests as high as 12° Baume, fermentation begins at about 77° F. and is raised during fermentation to 85° or 90° F. A temperature around 82° F. is best on the average as this is most conducive to the growth of yeast.

Where the vats are large and the syrup considerably diluted the temperature rises very quickly and must be moderated by passing a current of cold water through a coil of pipe on the bottom of the vat.

In the making of molasses mashes it must be remembered that every gallon of molasses will be diluted with about five gallons of water or other fermented liquid matter, and therefore 50 gallons of molasses wash will require a still capable of working up about 300 gallons. It is possible to distill four or five charges during the day of 12 hours and hence a still of 60 gallons will be capable of distilling the beer or wash made with 50 gallons of molasses. A still with a capacity of 100 gallons operating on wash having a strength of one gallon of molasses to five of water, will produce about 10 gallons of proof spirit from each charge; thus a 100 gallon still will make from 40 to 80 gallons of spirit in a day. With unskilled labor, however, it is impossible to get this rate of production and the best that can be done will be about four charges a day.

It may be suggested that in getting estimates on stills it is best to accompany the request with a statement of the character of the mash intended to be treated, the amount of raw materials intended to be used up, the charging capacity required, number of gallons of mash desired to be worked up every 12 hours.

Fermenting Raw Sugar. This is accomplished by dissolving the sugar in hot water, then diluting it, and then adding a ferment,—fermentation being aided by adding sulphuric acid to the diluted molasses, in the proportion of one-half to one pound of acid to every hundred pounds of pure sugar used.

The wash is pitched with compressed yeast in the proportion of 2½ to 8 per cent of the weight of the sugar used. The pitching temperature is from 77° to 79° F., and the period of fermentation is 48 hours.

Cane Sugar Molasses. Besides the molasses of the French beet sugar refineries, large quantities result from the manufacture of cane sugar in Jamaica and the West Indies. This is entirely employed for the distillation of *rum*. As the pure spirit of Jamaica is ne·er made from sugar, but always from molasses and skimmings, it is advisable to notice these two products, and, together with them, the exhausted wash commonly called *dunder*.

The molasses proceeding from the West Indian cane sugar contains crystallizable and uncrystallizable sugar, gluten, or albumen, and other organic matters which have escaped separation during the process of defecation and evaporation, together with saline matters and water. It therefore contains in itself all the elements necessary for fermentation, *i.e.*, sugar, water, and gluten, which latter substance, acting the part of a ferment, speedily establishes the process under certain conditions. *Skimmings* comprise the matters separated from the cane juice during the processes of defecation and evaporation. The scum of the clarifiers, precipitators, and evaporators, and the precipitates in both clarifiers and precipitators, together with a proportion of cane sugar mixed with the various scums and precipitates, and the "sweet-liquor" resulting from the washing of the boiling-pans, etc., all become mixed together in the skimming-receiver and are fermented under the name of " skimmings." They also contain the elements necessary for fermentation, and accordingly they very rapidly pass into a state of fermentation when left to themselves; but, in consequence of the glutinous matters being in excess of the sugar, this latter is speedily decomposed, and the second, or acetous fermentation, commences very frequently before the first is far advanced. *Dunder* is the fermented wash after it has undergone distillation, by which it has been deprived of the alcohol it contained. To be good, it should be light; clear, and slightly bitter; it should be quite free from acidity, and is always best when fresh. As it is discharged from the still, it runs into receivers placed on a lower level, from which it is pumped up when cool into the upper receivers, where it clarifies, and is then drawn down into the fermenting cisterns as required. Well-clarified dunder will keep for six weeks without any injury. Good dunder may be considered to be the liquor, or " wash," as it is termed, deprived by distillation of its alcohol, and much concentrated by the boiling it has been subjected to; whereby the substances it contains, as gluten, gum, oils, etc., have become, from repeated boilings, so concentrated as to render the liquid mass a highly aromatic compound. In this state it contains at least two of the elements necessary for fermentation, so that, on the addition of the third, viz., sugar, that process speedily commences.

The first operation is to clarify the mixture of molasses and skimmings previous to fermenting it. This is performed in a leaden receiver holding about 300 or 400 gallons. When the clarification is complete, the clear liquor is run into the fermenting vat, and there mixed with 100 or 200 gallons of water (hot, if possible), and well stirred. The mixture is then left to ferment. The great object that the distiller has in view in conducting the fermentation is to obtain the largest possible amount of spirit that the sugar employed will yield, and to take care that the loss by evaporation or acetification is reduced to a minimum. In order to ensure this, the following course should be adopted. The room in which the process is carried on must be kept as cool as it is possible in a tropical climate; say, 75° to 80° F.

Supposing that the fermenting vat has a capacity of 1000 gallons, the proportions of the different liquors run in would be 200 gallons of well-clarified skimmings, 50 gallons of molasses, and 100 gallons of clear dunder; they should be well mixed together. Fermentation speedily sets in, and 50 more gallons of molasses are then to be added, together with 200 gallons of water. When fermentation is thoroughly established, a further 400 gallons of dunder may be run in, and the whole well stirred up. Any scum thrown up during the process is immediately skimmed off. The temperature of the mass rises gradually until about 4° or 5° above that of the room itself. Should it rise too high, the next vat must be set up with more dunder and less water; if it keeps very low, and the action is sluggish, less must be used next time. No fermenting principle besides the gluten contained in the wash is required. The process usually occupies eight or ten days, but it may last much longer. The liquid now becomes clear, and should be immediately subjected to distillation to prevent acetous fermentation.

Sugar planters are accustomed to expect one gallon of proof rum for every gallon of molasses employed. On the supposition that ordinary molasses contains 65 parts of sugar, 32 parts of water, and three parts of organic matter and salts, and that, by careful fermentation and distillation, 33 parts of absolute alcohol may be obtained, we may then reckon upon 33 lbs. of spirit, or about four gallons, which is a yield of about 5⅔ gallons of rum, 30 per cent. over-proof, from 100 lbs. of such molasses.

The following process is described in Deerr's work on " Sugar and Sugar Cane."

" In Mauritius a more complicated process is

used; a barrel of about 50 gallons capacity is partly filled with molasses and water of density 1.10 and allowed to spontaneously ferment; sometimes a handful of oats or rice is placed in this preliminary fermentation. When attenuation is nearly complete more molasses is added until the contents of the cask are again of density 1.10 and again allowed to ferment. This process is repeated a third time; the contents of the barrel are then distributed between three or four tanks holding each about 500 gallons of wash of density 1.10 and 12 hours after fermentation has started here, one of these is used to pitch a tank of about 8,000 gallons capacity; a few gallons are left in the pitching tanks which are again filled up with wash of density 1.10 and the process repeated until the attenuations fall off, when a fresh start is made. This process is very similar to what obtains in modern distilleries save that the initial fermentation is adventitious.

" In Java and the East generally, a very different procedure is followed. In the first place a material known as Java, or Chinese, yeast is prepared from native formulæ; in Java, pieces of sugar cane are crushed along with certain aromatic herbs, amongst which galanga and garlic are always present, and the resulting extract made into a paste with rice meal; the paste is formed into strips, allowed to dry in the sun and then macerated with water and lemon juice; the pulpy mass obtained after standing for three days is separated from the water and made into small balls, rolled in rice straw and allowed to dry; these balls are known as Raggi or Java yeast. In the next step rice is boiled and spread out in a layer on plantain leaves and sprinkled over with Raggi, then packed in earthenware pots and left to stand for two days, at the end of which period the rice is converted into a semi-liquid mass; this material is termed Tapej and is used to excite fermentation in molasses wash. The wash is set up at a density of 25° Balling and afterwards the process is as usual. In this proceeding the starch in the rice is converted by means of certain micro-organisms *Chlamydomucor oryzae* into sugar and then forms a suitable habitat for the reproduction of yeasts which are probably present in the Raggi but may find their way into the Tapej from other sources. About 100 lbs. of rice are used to pitch 1,000 gallons of wash."

CHAPTER XI.

ALCOHOLOMETRY.

Alcoholmetry is the name given to a variety of methods of determining the quantity of absolute alcohol contained in spirituous liquors. It will readily be seen that a quick and accurate method of making such determinations is of the very utmost importance to those who are engaged in the liquor traffic, since the value of spirit depends entirely upon the percentage of alcohol which it contains. When alcoholic liquors consist of simple mixtures of alcohol and water, the test is a simple one, the exact percentage being readily deducible from the specific gravity of the liquor, because to a definite specific gravity belongs a definite content of alcohol; this is obtained either by means of the *specific gravity bottle*, or of hydrometers of various kinds, specially constructed.

All hydrometers comprise essentially a graduated stem of uniform diameter, a bulb forming a float and a counterpoise or ballast. The hydrometers may either be provided with a scale indicated on the neck or else with weights added to sink the hydrometer to a certain mark. The first instruments are called hydrometers of " constant immersion," the others, of "variable immersion."

At the latter end of the last century, a series of arduous experiments were conducted by Sir C. Blagden, at the instance of the British government, with a view to establishing a fixed proportion between the specific gravity of spirituous liquors and the quantity of absolute alcohol contained in them. The result of these experiments, after being carefully verified, led to the construction of a series of tables, reference to which gives at once the percentage of alcohol for any given number of degrees registered by the hydrometer; these tables are invariably sold with the instrument. They are also constructed to show the number of degrees over- or under-proof, corresponding to the hydrometric degrees. Other tables are obtainable which give the specific gravity corresponding to these numbers.

The measurement of the percentage of absolute alcohol in spirituous liquors is almost invariably expressed in volume rather than weight, owing to the fact that such liquors are always sold by volume. Nevertheless, the tables referred to above show the percentage of spirit both by volume and weight.

In the United States the standard liquor, known as *proof spirit*, contains 92.3 per cent. by weight and 94.9 per cent. by volume, of absolute alcohol; it has a specific gravity of .9186 at 60°F. A proof gallon contains by measurement 100 parts of alcohol and 81.5 parts of water. The strength and therefore the value of spirituous liquors is estimated according to the quantity by volume of an-

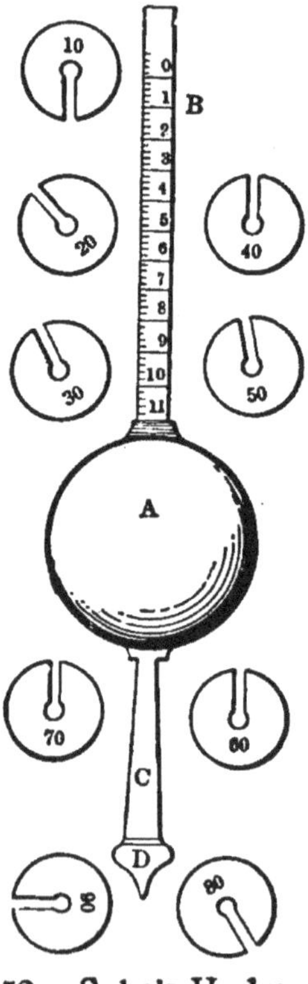

FIG. 52.—Syke's Hydrometer.

hydrous spirit contained in the liquor with reference to this standard. Thus the expression "20 *per cent. overproof*," "20 *per cent. underproof*," means that the liquor contains 20 volumes of water for every 10C volumes over or under this fixed quantity, and that in order to reduce the spirit to *proof*, 20 per cent. of water by volume, must be subtracted or added, as the case may be. Any hydrometer constructed for the measurement of liquids of less density than water may be employed. That known as " Syke's " is most commonly used for alcoholometric purposes. It is shown in Fig. 52 and consists of a spherical brass ball *A*, to which is fixed two stems; the upper one *B* is also of brass, flat, and about 3½ in. in length; it is divided into ten parts, each being subdivided into five, and the whole being numbered as shown in the figure. The lower stem *C* is conical, and slightly more than an inch long; it terminates in a weighted bulb *D*. A series of circular weights, of the form shown in the figure, accompany the instrument; these are slipped upon the top of the lower stem *C*, and allowed to slip down until they rest upon the bulb *D*. The instrument is used in the following way: It is submerged in the liquor to be tested until the whole of the upper stem is under the surface, and an idea is thus gained of the weight that will be required to *partly* submerge the stem. This weight is added, and the hydrometer again placed in the liquor. The figure on the scale to which the instrument has sunk when at rest is now observed, and added to the number on the weight used, the sum giving, by reference to the tables, the percentage by volume of absolute alcohol above or below the standard quantity.

In exact estimations, the temperature of the liquor tested must be carefully registered, and the necessary corrections made. In Jones's hydrometer, which is an improvement upon Syke's, a small spirit thermometer is attached to the bulb, and by noting the temperature of the liquor at the time of the experiment, and referring to the tables accompanying the instrument, the strength is found at once without the need of calculation.

Dica's hydrometer is very similar to Jones's instrument above described. It is of copper, has a stem fitted to receive brass poises, a thermometer, a graduated scale, etc.

In Europe, Gay-Lussac's hydrometer and tables are chiefly used for alcoholometric testing. This instrument is precisely similar in construction to those of Twaddle and Baume. On the scale, zero is obtained by placing it in pure distilled water at 59° F., and the highest mark, or 100, by placing it in pure alcohol at the same temperature, the intermediate space being divided into 100 equal divisions, each representing one per cent. of absolute alcohol. The correction for temperature, as in the above cases, is included in the reference tables.

Another hydrometer, used in France for alcoholometric determinations, is Cartier's. In form it is precisely similar to Baume's hydrometer. Zero is the same in both instruments, but the point marked 30° in Cartier's is marked 32° in Baume's, the degrees of the latter being thus diminished in the proportion of 15 or 16. Cartier's hydrometer is only used for liquids lighter than water.

The alcoholmeter of Tralles is the official instrument for testing alcoholic liquors in the U. S. but the instrument which is most generally used both here and abroad is that of Beaumé. There are two instruments bearing Beaumé's name, one for liquids lighter than water, the other for those which are heavier. All hydrometers, alcoholmeters and saccharometers work on the same principle, though they are each differently graduated for the particular work to be done and the details of the measuring process are slightly different. All these instruments are provided with tables whereby their readings may be corrected and the specific gravity of the liquid determined.

The above hydrometric methods can be safely employed only when the spirit tested contains a very small amount of solid matter, since, when such matter is contained in the liquor in quantity, the density alone cannot possibly afford a correct indication of its richness in alcohol. Many methods have been proposed for the estimation of alcohol in liquor, containing saccharine coloring and extractive matters, either in solution or suspension. Undoubtedly the most accurate of these, though at the same time the most tedious, is to subject the liquor to a process of distillation by which a mixture of pure alcohol and water is obtained as the distillate. This mixture is carefully tested with the hydrometer, and the percentage of alcohol in it determined by reference to the tables as above described; from this quantity and the volume of the original liquor employed the percentage by volume of alcohol in that liquor is readily found. The condensing arrangement must be kept perfectly cool, if possible in a refrigerator, as the alcohol in the distillate is very liable to be lost by re-evaporation. When great accuracy is desired, and time is at the operator's disposal, the above method is preferable to all others.

It is performed in the following manner: Three hundred parts of the liquor to be examined are placed in a small still, or retort, and exactly one-third of this quantity is distilled over. A graduated glass tube is used as the receiver, in order that the correct volume may be drawn over without error. The alcoholic richness of the distillate is then determined by any of the above methods, and the result is divided by three, which gives at once the percentage of alcohol in the original liquor. The strength at proof may be calculated from this in the ordinary way.

If the liquor be acid, it must be neutralized with carbonate of soda before being submitted to distillation. From eight to ten per cent. of common salt must be added, in order to raise the boiling point, so that the whole of the spirit may pass over before it has reached the required measure. In the case of the stronger wines it is advisable to distil over 150 parts and divide by two instead of three. If the liquor be stronger than 25 per cent. by volume of alcohol, or above 52 to 54 per cent. under-proof, an equal volume of water should be added to the liquid in the still, and a quantity distilled over equal to that of the sample tested, when the alcoholic strength of the distillate gives, without calculation, the correct strength required. If the liquor be stronger than 48 to 50 per cent. under-proof, three times its volume of water must be added, and the process must be continued until the volume of the distillate is twice that of the sample originally taken. In each case the proportionate quantity of common salt must be added.

For the estimation of alcohol in wines, liquors, etc., the following method may be employed: A measuring flask is filled up to a mark on its neck with the liquor under examination, which is then transferred to a retort; the flask must be carefully rinsed out with distilled water, and the rinsings added to the liquor in the retort. About two-thirds are then drawn over into the same measuring flask, and made up to its previous bulk with distilled water, at the same temperature as that of the sample before distillation. The strength is then determined by means of Syke's hydrometer, and this, if under-proof, deducted from 100, gives the true percentage of proof-spirit in the wine.

A quick, if not always very exact, method consists in determining the point at which the liquor boils. The boiling point of absolute alcohol being once determined, it is obvious that the more it is diluted with water the nearer will the boiling point of the mixture approach that of water; moreover, it has been proved that the presence of saccharine and other solid matters has but an almost inappreciable effect upon this point. Field's alcoholometer, since improved by Ure, is based upon this principle. It is shown in Fig. 53, and consists, roughly speaking, of a cylindrical vessel *A*, to contain the

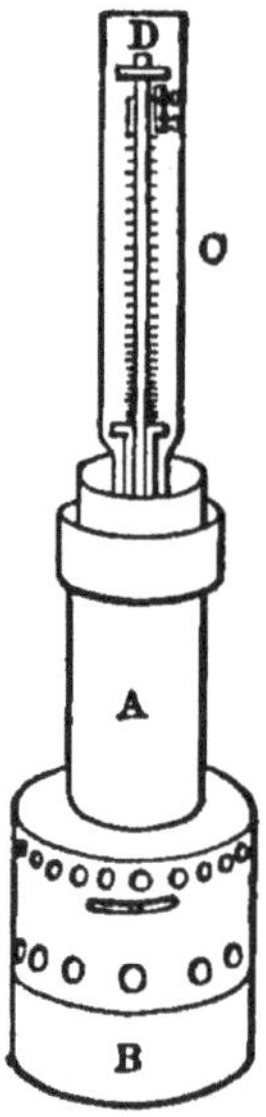

FIG. 53.—Field's Alcoholometer.

spirit; this vessel is heated from beneath by a spirit lamp, which fits into the case *B*. A delicate thermometer *C*, the bulb of which is introduced into the spirit, is attached to a scale divided into 100 divisions, of which each represents one degree over- or under-proof. This method is liable to several small sources of error, but when a great many determinations have to be made, and speed is an object rather than extreme accuracy, this instrument becomes exceedingly useful. It does not answer well with spirits *above* proof, because the variation in their boiling points are so slight as not to be easily observed with accuracy. But for liquors under-proof, and especially for wines, beer, and other fermented liquors, it gives results closely approximating to those obtained by distillation, and quite accurate enough for all ordinary purposes. Strong liquors should therefore be tested with twice their bulk, and commercial spirits with an equal bulk, of water, the result obtained being multiplied by two or three, as the case may be.

Another very expeditious, but somewhat rough, method was invented by Geisler. It consists in measuring the tension of the vapor of the spirit, by causing it to raise a column of mercury in a closed tube. The very simple apparatus is shown in Fig. 54. *A* is a small glass bulb, fitted with a narrow tube and stop-cock. This vessel is completely filled with the spirit, and is then screwed upon a long, narrow tube *B*, bent at one end and containing mercury. This tube is attached to a graduated scale showing the percentage of absolute alcohol above or below proof. To make the test the cock is opened, and the bulb, together with the lower part of the tube, is immersed in boiling water, which gradually raises the spirit to its boiling-point. When this is reached, the vapor forces the mercury up the tube, and, when stationary, the degree on the scale to which it has ascended gives directly the percentage of alcohol.

Another method, which is not to be relied on for very weak liquors, but which answers well for cordials, wines, and strong ales, is that known as Brande's method. The liquor is poured into a long, narrow glass tube, graduated centesimally,

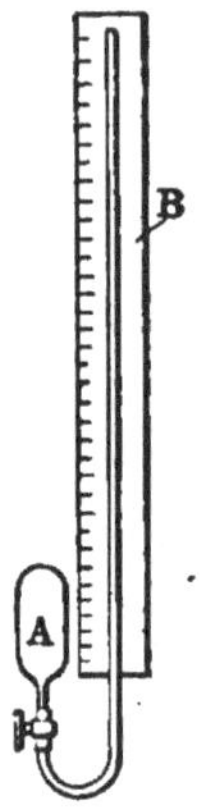

Fig. 54.—Geisler's Apparatus.

until it is half-filled. About 12 or 15 per cent. of subacetate of lead, or finely powdered litharge, is then added, and the whole is shaken until all the color is destroyed. Powdered anhydrous carbonate of potash is next added until it sinks undissolved in the tube, even after prolonged agitation. The tube is then allowed to rest, when the alcohol is observed to float upon the surface of the water in a well-defined layer. The quantity read off on the scale of the tube and doubled, gives the percentage by volume of alcohol in the original liquid. The whole operation may be performed in about five minutes, and furnishes reliable approximative results. In many cases it is necessary to add the lead salt for the purpose of decolorizing the liquid.

For the investigation of the amount of sugar in, or the concentration of the mash, or beer, a specially scaled hydrometer is used which is termed a saccharometer. Sugar possesses a higher degree of specific gravity than water, and hence it follows that the greater the amount of sugar in the mash the higher will be the specific gravity. The less the hydrometer sinks into the fluid the greater the amount of sugar present. Saccharometers are provided with thermometers whereby the reading may be corrected to a standard temperature, usually 59° F. The saccharometer is correct for solutions containing sugar alone but it is only approximately correct for mash liquor which contains a variety of other matters in variable quantities.

It is a prime necessity that the distiller should be able to determine if the mash has been completely saccharified by the malt. For this purpose a solution of iodine is used. Iodine gives to starch a blue color, If the starch however, has been completely changed into sugar there will either be no discoloration or the filtered mash liquid which is at first a yellowish red becomes blue, then violet, and at last red.

Determination of the Purity of Alcohols. While the knowledge of the amount of alcohol contained in a liquid is of great practical utility, this does not give any idea of the impurities present.

An alcohol of 100 degrees or an absolute alcohol, may contain numerious impurities which may greatly affect its quality. It is therefore necessary in addition to analyze the purity of the alcohol.

In commercial practice there are certain simple processes which will give a basis by which to determine the impurities left after distillation and rectification. These processes are largely empirical They are based on the perception of the senses and are consequently of an entirely relative degree of precision. Nevertheless, when made by a practical expert, the operation may give very useful preliminary indications.

This test is made in a glass of special shape broad at the bottom and narrowing at the top in order to concentrate the aroma of the product. Ordinary brandies are tested undiluted. Commercial alcohols, of about 95 degrees must be diluted with water to a maximum of 30 degrees. Otherwise the burning tang of the alcohol would preclude any delicacy of perception and allow impurities to pass unnoticed.

The operation is begun by examination by sense of smell. The glass is half filled with the liquid diluted with one half of pure water. The glass is covered with one hand and shaken violently for a few seconds. Immediately upon uncovering it, the quality of the alcoholic vapors may be ascertained by their odor.

For the examination by sense of taste, the operator rinses his mouth for a moment with the liquid itself. The taste of ethyl alcohol is fairly transient;—it disappears quickly allowing the taste of the accompanying foreign matter to be perceived almost immediately afterward. With a little practice this test enables one to distinguish by their flavor the primal origin of alcohols and to judge of their purity. Some professionals succeed by training in arriving at high degree of skill in the art of tasting alcohol as it should be done.

In order to determine the purity of alcohol there are besides chemical tests used by the trade. These tests, which consist in characterizing and measuring separately the impurities which alcohol may contain, such as acids, ethers, aldehydes, bases, etc., belong exclusively to analytical chemistry; they are extremely delicate and complicated. We will not venture to touch upon them here.

One of the simplest tests for purity is that of Barbet. This is based upon the time taken to discolor a solution of permanganate of potash under the action of the tested alcohol. It is not only very rapid but in general more practical than other tests. It allows the aggregate of the impurities contained in an alcohol to be ascertained in a single operation.

The permanganate solution used is very weak (0. gr. 200 of salt), and of a violet-red color.

The technique of the proceeding is as follows: 50 cubic centimeters of the alcohol to be tested are placed in a glass vessel the temperature of which is maintained at 64.40°F. 2 cubic centimeters of the permanganate solution are abruptly added and the time noted to within a second. Discoloration is awaited and as soon as it takes place the time is again noted. The total discoloration of the permanganate is not very marked and passes through intermediate stages; therefore it is preferable not to await complete discoloration but to stop at a pale salmon tint, which tint may be comparatively fixed by a sample of colored liquid (say a solution of fuchsine and chromate of potash).

The comparative times of discoloration obtained by M. Barbet with various commercial alcohols, are as follows:

Pure alcohol	43 min.	30 sec.
Extra fine alcohol	5 "	30 "
Semi fine alcohol	5 "	10 "
Medium flavor alcohol (first running)	5 "	5 "
Mediocre alcohol	5 "	11 "
Medium flavor alcohol (last running)	2 "	12 "

CHAPTER XII.

Distilling Plants: Their General Arrangement and Equipment.

When we look at the manufactories of to-day with their complicated machinery, their extensive equipment, their great boilers, and engines and their hundreds of employees, we are liable to forget that good work was turned out by our ancestors, with equipment of extreme simplicity and that to-day while there are, for instance, thousands of wood-working mills, complete in every detail and covering under a multitude of roofs every variety of complicated and perfected wood-working machinery, yet there are many more thousands of small plants, comprising a portable boiler, fed with refuse, a small engine and a few saws which are making money for the owners and doing the work of the world.

The reader therefore, must be warned against any feeling of discouragement because of the cost and complicated perfection of elaborate distilling plants. Where the business is to be entered into on a large scale, to take the products from a considerable section of country and turn them into alcohol to compete in the great markets, the best of apparatus and equipment is not too good, but the person contemplating the mere manufacture of alcohol on a small scale, to serve only a small section, must remember that distillation is really a very simple matter, for years practiced with a most rudimentary apparatus and still so practiced in the country districts particularly in the South.

This is well illustrated by the fact that an illicit distiller confined in one of the North Carolina penitentiaries for transgressing the revenue laws, was able while in durance, to continue his operations unknown to the prison authorities, his plant consisting of a few buckets, and a still whose body was a tin kettle, a few pieces of pipe and a worm which he had bent himself. This example is not given as encouragement to illicit or "blockade" distilling but merely to show vividly how simple the rudimentary apparatus really is.

The simplest regular plants, those of the South for instance, comprise a building of rough lumber some thirty feet by twelve wide, with a wooden floor on which the fermenting vats rest and an earthern floor immediately in front of the still and furnace. This is to permit the fires being drawn when the charge has been exhausted in the boiler. The still is of the fire-heated, intermittent variety, such as described on page 35. It consists of a brick furnace or oven, large enough to burn ordinary cord wood and supporting a copper boiler of fifteen or twenty gallons capacity. On top of this is a copper "head" with the usual goose neck, from which a copper pipe leads to a closed and locked barrel containing raw spirits, this barrel acting on the principle of the condensing chamber shown in the still in Fig. 8 From the upper part of this barrel, which acts as a concentrator, the vapors pass to a copper worm immersed in a tub of cold water. Here the vapors are condensed and pass by a pipe to a small room, containing a locked receiving tank. This room is kept locked and is under the immediate charge of the Government officer in charge of the still, or, in the case of alcohol intended for de-naturing, the alcohol would pass to a locked tank from whence it would be taken and de-natured under the charge of the proper Government officer.

The fermenting vats may be six or more in number so as to allow the mash in each tank to be at a different stage of fermentation. A hand pump is used for pumping the contents of any of the tanks into the boiler or the still. A hand pump is also provided for supplying water to the vats and condensers.

In connection with the distilling and fermenting building there are small buildings for storing the grain, malt, etc., for the storage of the alcohol and for the keeping of the various books, records, and stamps required by law. Such plants as these are located adjacent to a good clear spring or even a small brook, and preferably in a position convenient to the carriage of materials and the transportation of the whiskey or other liquor produced.

The buildings are of the cheapest construction

and arranged in the manner which compels the least labor in filling the mash vats and turning the contents into spirits. There are no special mash coolers, no complicated stirrers. The "beer" as the fermented mash is called is stirred by a paddle in the hands of a strong negro and the mash is mixed and fermented by rule of thumb, without the use of any scientific appliances. Primitive, as it is, however, those small plants in certain sections of the country make money for their proprieters and serve a large number of customers. The spirits so produced are low grade, fiery and rough in taste, but the point is that alcohol may be and is so produced.

Between these simple beginnings and the elaborate plants of big distilleries there is a wide range, so wide that it is impossible within the limits of this book to go into detail. The makers of distilling apparatus furnish all grades of stills and to those contemplating erecting a plant it is suggested that their best course is to communicate with such manufacturers, giving the circumstances of the case, the particular product to be worked and the capacity desired. The object of this book is to give an understanding of the processes of distillation and of this chapter to give a general idea of the arrangement of a number of typical distilling plants, suitable for various kinds of work.

That the simple, direct-heated pot still such as referred to above, used for fifteen hundred years and over, is still used is largely due to the simplicity of its construction and operation, but its capacity is small, and its operating expense relatively heavy. It is still used for making liquors, but for industrial purposes it has been entirely superceded by concentrating and rectifying stills. A simple form of the latter is found in the still shown in Fig. 11 and in the distilling apparatus of Adam (Fig. 9).

Originally all stills were heated by direct contact with fire. This was open to a serious objection, namely, that the mash if thick was liable to be scorched. Stirring devices were used by Pistorious but these required constant attention. As a consequence, direct firing gave place to heating by steam, by which not only was scorching of the wash avoided but much greater certainty of operation was attained.

The steam may be used to simply heat the boiler, thus taking the place of the direct heat of the fire, but it is far better in every way to admit the steam directly to the mash as in the Coffey still, Fig. 18, and all modern stills. It is possible to apply this principle to all compound stills, but the best results with greatest economy of fuel are, of course, gotten from the plate or column stills especially constructed for steam. In order to get the best results it is necessary that the entry of steam be regulated so that there may be absolute uniformity of flow. A convenient form of regulator is that invented by Savalle, and described on page 70, but there are a number of other forms on the market each one having its special advantages.

It will be seen then that while the simple pot still, fire-heated, may be used, the practical plant for the fermentation of industrial alcohol should have a modern continuous still and rectifier and a boiler for generating the necessary steam for it and for the operations of mashing and fermenting.

The Fermenting Room.

The fermenting room has three main requirements for successful commercial distillation. It must allow a uniform temperature to be maintained in the vats; it must have thorough ventilation without any draftiness, and it must be absolutely clean. It should have also plenty of light so that it may be thoroughly inspected. It is true that in the primitive plants all these requisites were violated, but there is no reason for this. The first cost is but little added to by building with these requisites in mind and it is far more profitable in the long run; and it is only by the elimination of the bacteria which are inimical to proper fermentation that the fermenting operation can be performed with any certainty.

For the regulation of the temperature reliance may be had on stoves or heaters, or on special mash heaters and coolers by which the temperature of the mash in the tubs may itself be controlled without reference to the temperature of the fermenting room. When, however, no special and adequate heating means is provided, the walls should be double with an air space between and the doors and windows should either be also double or limited in number.

To ensure good ventilation and plenty of space above the vats wherein to work or install suitable vatting machinery, the walls should be at least twelve feet in height. Outlet openings should be formed around the base of the room leading to the outer air and closed by controllable shutters. These are to allow the escape of the carbonic acid gas evolved during fermentation. These should be most carefully constructed, however, to pervent drafts.

The walls and floor of the fermenting rooms should be so made that they may be easily washed down and kept clean. Concrete floors are excellent for this purpose and the walls also may be faced with concrete or cement covered with a coating composed of a mixture of asphalt and coal tar. This mixture may be also applied to plaster walls with good results.

The fermenting vats, as before stated, are made of wood for small plants, and of iron for larger plants, and are usually from three and a half to four and a half feet in height. After the chief feremnting period, it is necessary that the temperature of the mash be prevented from rising beyond 86° F. and to that end movable cooling tubes, coils and stirrers are used. These consist of parallel frames made up of tubes, preferably of copper, through which cold water is passed and which are moved about in the vat, either vertically or rotatively. There must be space above the vats, therefore, for the introduction and removal of these cooling frames, and for the gearing whereby they are driven.

As previously stated, mashes to-day are mostly prepared by steaming and disintegrating in a mash cooker of the type shown in Figs. 1 and 41 or in Henze steamers, from which the mash is blown into the preparatory mash vat, where it is stirred and brought to the proper temperature for fermentation. A convenient arrangement of mash cooker, coolers, pump and vats is shown in Fig. 1. Where Henze steamers are used they are arranged in batteries, the blow-off pipes being connected to the preparatory mash vats. These are preferably provided with water cooled stirrers consisting of a frame of straight and vertical tubes mounted on a tubular arm projecting from a tubular shaft, and rotated in a horizontal plane wi hin the closed mash vats, by suitable gears. The rotation of the arm stirs and automatically mixes the mash while cooling it. Another form of cooler is shown diagrammatically in Fig. 4.

Whatever form of cooling apparatus is used, attention should be paid to the ease with which the stirrers or tubes can be kept clean, and to the strength of the apparatus, gears, etc. Concen-

FIG. 56.—Grain Distillery. Capacity

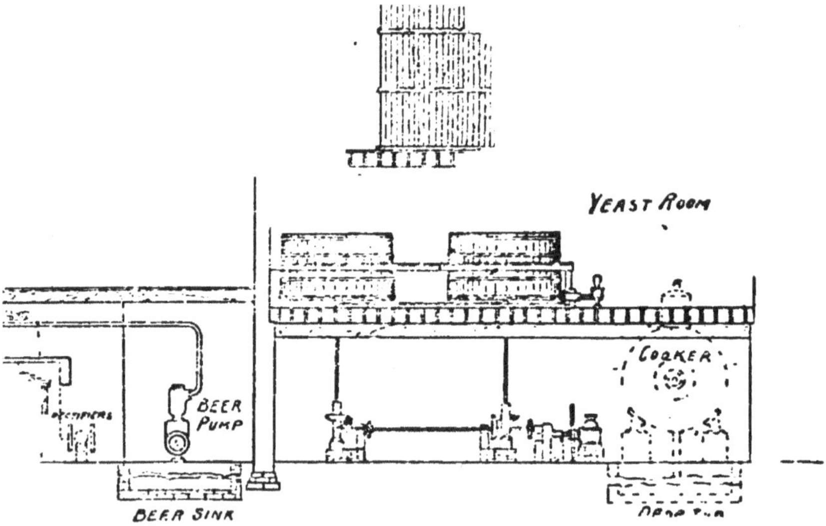

600 Bushels per day.

(To face page 198)

trated or thick mashes require that the stirrers be of massive construction, capable of being rapidly rotated in the liquid.

In preparatory mash vats for use with concentrated mashes, means must also be provided for clearing the mash. These mash cleaners and husk removers usually form part of, or are attached to the vat itself and are driven by gearing from the main shaft carrying power to the mashing room.

A good idea of the general arrangement and correlation of the various apparatus of a plant may be gathered from the sectional view of a grain distillery shown in Fig. 55. It will be seen from this that the mashing apparatus, steamers and mixers are located on the several floors of one building and in such relation to each other that the several operations of saccharifying are carried on in a continuous movement of mash towards the fermenting vats.

Adjoining the fermenting vat room is a section of the plant given up to the manufacture of pure yeast and this and the fermenting rooms are level with the ground, have solid walls whereby a uniform temperature is obtained, and plenty of space for proper ventilation of the vats. A gallery traverses the room about midway the height of the vats so that convenient access may be had to them. The distilling room is high enough to allow for the setting of the various columns, separators and condensers at their proper heights relative to each other, and

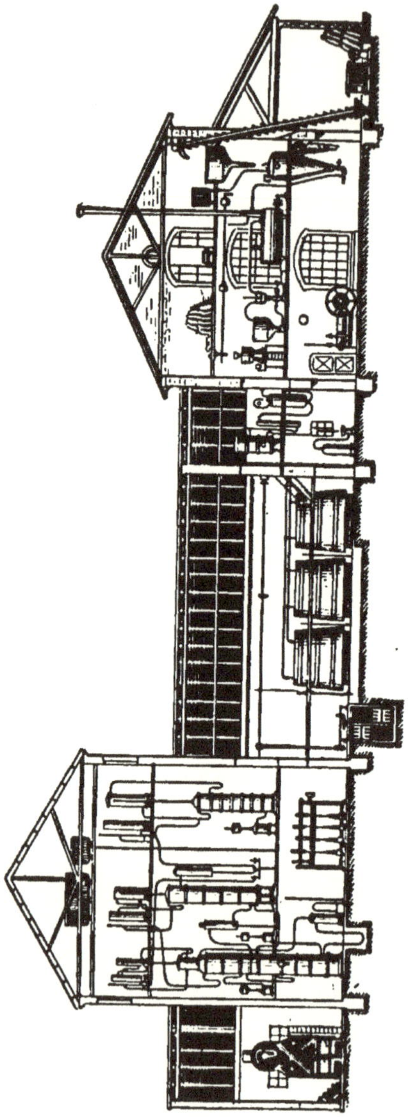

FIG. 55.—Continuous Grain Alcohol Distillery—Barbet's System.

should be so arranged as to its several floors or stages that access to the various pipes and apparatus may be easily had. The steam generator for the column is located in an adjacent room.

In addition to this there should be a malt house for the preparation of malt, located conveniently to the saccharifying building; an engine and boiler room so placed that power may be conveniently transferred to the mixers, stirrers and pumps and to generate steam for the Henze boilers; while adjacent to the distilling building should be the storage tanks and de-naturing department.

Another arrangement of apparatus for a grain distillery with a capacity of 2500 bushels per day is illustrated in Fig. 56. This plant was erected by the Vulcan Copper Works Co., and includes separate stills for gin, alcohol, and rye whiskey, as well as a spirit rectifying column.

The milling and grain mixing departments, the yeast room and the fermenting room are arranged on the several floors of one building. in the basement of which is located the vacuum cooker and drop tub and coolers described on page 11 from which the mash is pumped into the fermenting tubs.

The second section of the building contains the distilling apparatus, storage tanks, charcoal rectifiers and spirit rectifying apparatus, while the third. section of the building comprises the boiler house and engine room.

In Fig. 57 is shown a view of a small plant for the distillation of beets, the figure giving a good

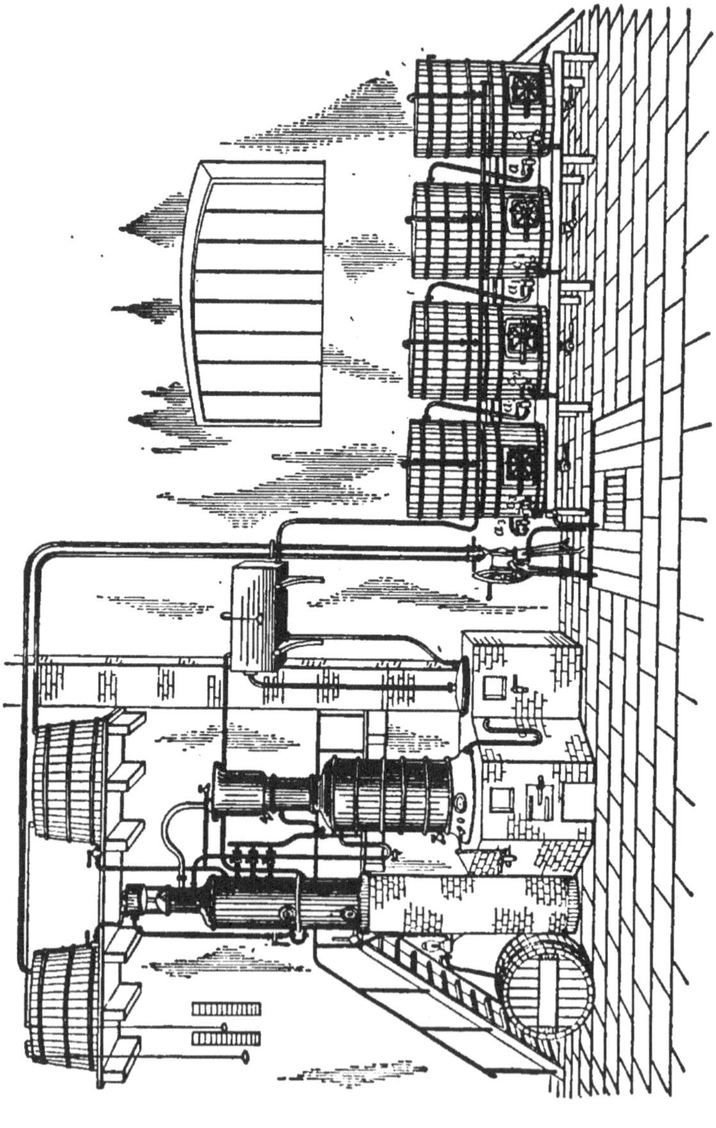

Fig. 57—Small Beet Distillery.

idea of the arrangement of the diffusion battery in relation to the still and rectifier. The juice from the diffusion battery is pumped into the overhead tanks from which it descends into a dephlegmator and from thence into the still, the vapors from the still passing into the rectifier. The still is a direct, fire-heated still and adjacent to the still is a water heater from which the water passes to the hot water reservoir located above and to one side of the diffusion vats.

A large plant for the distillation of beets is shown in the Section Fig. 58. The beets from the beet silos are carried to suitable washing machines, *A*, see Chapter VII, in which they are thoroughly cleaned of dirt and gravel. From the washers they are lifted by a conveyor *B* to a distributor *C* by which they are conveyed to the cutters or slicers. These consist of horizontal apertured plates revolving at a high speed, and carry knives which plane off slices from the beets. These drop through the apertures of the plate and are conveyed to the diffusion batteries, as by a movable chute *D* oscillated with a jigging motion through suitable gearing.

The diffusers *F* should be arranged so that small trucks may be driven beneath them to receive the spent slices and carry them to the spent beet silos. *U* indicates a gauging tank into which the juice runs from the diffusers. From thence it passes to coolers (not seen) and thence to the fermentation tanks *G*. *R* indicates

Fig. 58.—Large Beet Distillery.

a small engine for driving the beet slicers and *S* a battery of pumps whereby the wash may be forced up into the reservoir *I* from which the wash descends into the still *K*. *H* and *J* are reservoirs for hot and cold water respectively.

From the distilling column *K* the phlegm or raw spirit passes to the phlegm tank *L* from which it is drawn as desired into the rectifying column *M*, thence into the coolers and condensers and thence into the alcohol tanks *N*.

On the other side of the building as indicated by the chimney is the boiler for generating the motive power for the plant and for supplying the steam necessary for the distilling and rectifying columns and the hot water for the diffusion batteries. The boiler should be very capacious and it would be well to have two, one in reserve.

If possible, advantage should be taken of the natural slope of the ground so that the trucks bringing beets from the silo to the washer and carrying the spent beets away may roll downward by their own weight. The silos for the spent beets should be excavated from the ground and the trucks be constructed to tip their contents into these pits. The natural slope of the bottom of these pits should drain away the water and means be provided whereby carts can load with the spent beets to carry them away.

The spent liquors should flow off into ponds from which they may be drawn away to fertilize land.

A very convenient method of carrying beets from the silos to the washing machine is by means of a narrow canal of rapidly flowing water, flowing between the silos and entering the washing machines. Beets pitched into this stream are carried along by the current to the washers and at the same time undergo a preliminary washing. By laying out a system of channels throughout the beet yard the labor of handling is reduced to a minimum. These channels may be covered by boards on which the beets may be piled. These may be lifted and the beets thereon dumped into the stream.

A plant for the distillation of potatoes would be arranged very much after the plan of the grain distillery heretofore described except that it would have to be provided with apparatus for washing the potatoes and removing stones and adhering clods of earth. These washers, as put on the market, comprise a slotted rotating drum, which tumbles the potatoes about and loosens the dirt. When they escape from the drum they enter a washing trough where they are stirred about by revolving blades and acted upon by a swift current of water. The trough should be about two feet long to properly wash the potatoes. They are then lifted by an elevator to the mouth of the Henze pulpers (see Fig. 2) or the vacuum cookers see Fig. 1).

It is of advantage that the washing apparatus be so located that the potatoes as they are received

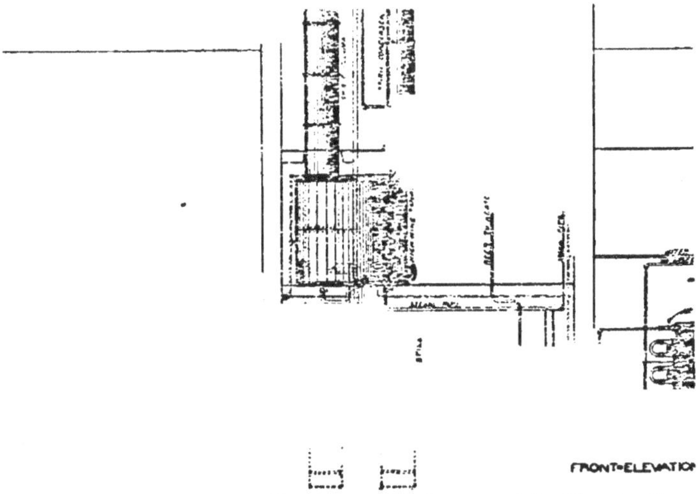

Fig. 59.—Molasses Distillery. Cap

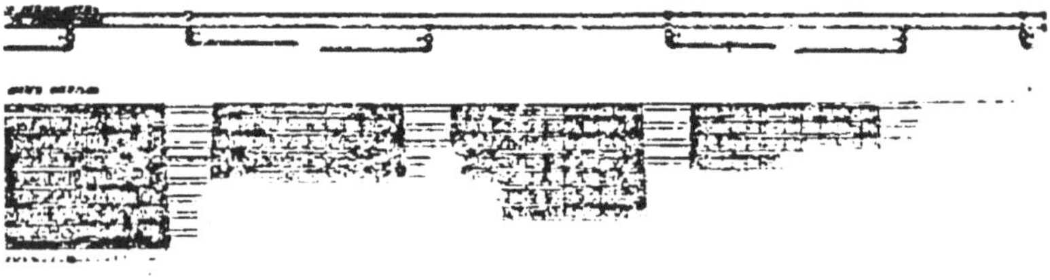

acity 2,500 gallons per day. (To face

may be shoveled into it immediately. The scale for weighing the potatoes as they are brought in should be so located that the manager may attend to the weighing without having to leave the distillery. This and other like details may seem of small moment but it is care in such details which conduces to the success of a plant. As before stated in describing a beet distillery, advantage should be taken of the lay of the land in laying out the plant so that the spent pulp may be easily disposed of, the spent wash carried away, and the finished product conveniently handled.

In Fig. 59, is shown a plant for distilling molasses, designed by the Vulcan Copper Works, before referred to, and erected for the Rio Tamposo Sugar Co., of Tamposo, S. L. P., Mexico.

The molasses as before explained at page 164 being too concentrated, is first pumped into the steam mixing tank on the ground floor of the distilling building. Here it is diluted and heated, mixed with sulphuric acid and pumped into the long ranges of cooling pipes, located along the fermenting room and built on the principle shown in Fig. 4. Here it is further diluted and yeast is added. From the fermenting tubs the molasses beer is pumped into the beer heater and thence into such a still as is shown in Fig. 32.

In addition to this the plant contains a rectifying apparatus for the high wines produced by the beer still, comprising a spirit still, charged from a high wine tank, a rectifying column, separator, and tubular condenser from which the rectified spirit is carried to the storage tanks.

Cane sugar distilleries are practically arranged the same as the molasses distillery above described. The cane is crushed between the rolls of cane crushers on the receiving floor and is then strained to remove the "begasse." The clarified juice is then pumped up to the mixing

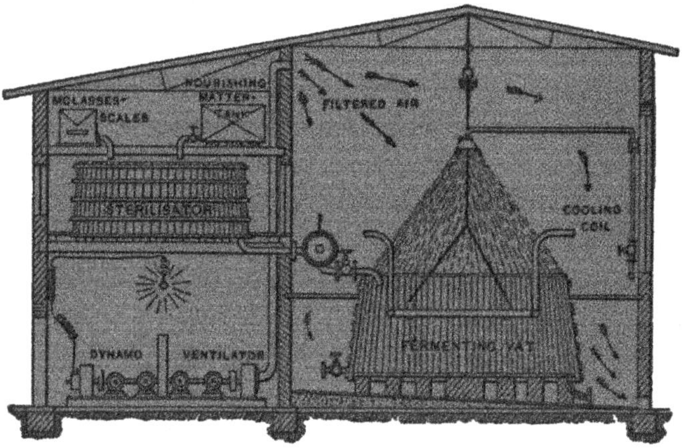

Fig. 60.—Molasses Fermenting House.

tanks. In these the molasses is mixed with spent wash from other fermentations or with water, after which it is acidified and flows to the fermenting vats. The fermenting house should be provided with means for forcing in filtered air and for ventilating, as molasses wash is very sensitive to change in temperature and very liable to become contaminated by injurious ferments. (See Fig. 60).

Above each vat should be a cooling coil capable of being lowered into the vat and a water spraying pipe, whereby the mash may be diluted when desired. From the vats, the wash is pumped to the distilling and rectifying columns. In Jamaica the still shown in Fig. 37, is largely used, as also the Coffey still, Fig. 18.

It is very often not profitable to distill spirit from molasses or sugar cane directly at the sugar factories, there being no market on the spot and transportation of the spirit in casks being very costly and difficult, not only because of the lack of transporting means but because the tropical climate tends to warp the empty casks. Transportation of the molasses in casks to a distillery is likewise open to objections of cost and the action of the hot sun in fermenting the molasses and bursting the cask.

Barbet has suggested a way out of the difficulty. This consists in boiling the molasses in vacuo, and then running it into molds lined with sheets of paper. These are set by dipping in cold water. When set the loaves wrapped in their paper coverings are as easily handled as sugar loaves. There is no dead weight nor any " empties " to be returned as in the case of casks. The molasses is in a most concentrated form and this makes for economy in freight. There is no risk of deterioration and the loaves may be stored in an ordinary warehouse. This method allows the distillery to be located at centers of transportation or at seaports, while the sugar factories are on the plantation.

Care should be taken in selecting the site for a distillery that an abundance of pure water may be supplied. The purer the water the better, and where water is not pure, purifying apparatus should be provided. The coolness of the water is a factor which must be taken into consideration. The greater amount of water will be used for cooling, and it follows then that the cooler the water the less of it will have to be used.

The horse-power of the engines used in driving the distilling apparatus varies, of course, with the capacity of the still, the average being between 6 H.P. and 30 H.P., for plants having fermenting vats of capacities ranging between two hundred and fifty, and twelve hundred gallons.

It must not be forgotten that the coal consumption of a plant depends upon the economy of heating means in the distilling apparatus, the perfection with which the heat of the vapors is used to heat the wash, the perfection of the boiler grates and the method of firing. These latter matters should be obvious to any distiller, but it is in economy in little things that the successful operation of a plant resides.

Nothing is more surprising than the difference in the coal consumption of different distilleries. Some use a third more than others. This is caused by poor coal, by poor firing, by poor boilers, by hard water, or by poor distilling equipment. With regard to the latter this word of advice may be given: The greater the number of plates in the distilling column, the less the coal consumed per gallon of alcohol produced. It must, however, be taken into account that a large number of plates in a column means a column of considerable height and that in turn means a correspondingly tall still house and increased first cost. Hence it is more economical to use the best forms of traps on the plates and fewer plates, and the best forms of these traps as pointed out in Chapter III, are those wherein the largest quantity of vapor in a finely divided state may come into contact with the greatest number of liquid particles.

In conclusion it may be said that dirt, neglect, carelessness and a too great desire for economy in first cost are all factors in lowering the economical productiveness as well in a distillery as in other manufacturing plants.

CHAPTER XIII.

De-natured Alcohol and De-naturing Formulæ

The uses of alcohol are very numerous and varied, the principal being, of course, for the production of all alcoholic liquors such as brandy, gin, rum, whiskey, liquors, etc.; that distilled from grain is almost entirely consumed in the manufacture of whiskey, gin; and British brandy. In the arts, strong alcohol is employed by the perfumers and makers of essences for dissolving essential oils, soaps, etc., and for extracting the odor of flowers and plants; by the varnish-makers for dissolving resins; by photographers in the preparation of collodion; by the pharmaceutists in the preparation of tinctures and other valuable medicaments; by chemists in many analytical operations, and in the manufacture of numerous preparations; by instrument makers in the manufacture of delicate thermometers; by the anatomist and naturalist as an antiseptic; and in medicine, both in a concentrated form (rectified spirit), and diluted (proof spirit, brandy, etc.), as a stimulant, tonic, or irritant, and for various applications as a remedy. It is largely consumed in the manufacture of vinegar; and in the form of methylated spirit it is used in lamps for producing heat. It has, in fact, been employed for a multitude of purposes which it is almost impossible to enumerate.

The common form of alcohol known as " de-natured spirit " consists of alcohol to which one tenth of its volume of wood alcohol, or other de-naturizing agents has been added, for the purpose of rendering the mixture undrinkable through its offensive odor and taste. Methylated spirit being sold tax free, may be applied by chemical manufacturers, varnish makers, and many others, to a variety of uses, to which, from its greater cost, duty-paid spirit is commercially inapplicable. Its use, however, in the preparation of tinctures, sweet spirits of nitre, etc., has been prohibited by law. It has often been attempted to separate the wood spirit from the alcohol, and thus to obtain pure alcohol from the mixture, but always unsuccessfully, as, although the former boils at a lower temperature than the latter, when boiled they both distil over together, owing probably to the difference of their vapor densities.

It is Germany which has led the way in the manufacture and use of " de-natured " alcohol or " spiritus," as it is there known. Germany has no natural gas or oil wells, and gasoline and kerosene are not produced there, hence the necessity of using some other form of liquid fuel. This fuel—in many ways better than any petroleum product—was found in alcohol. The sandy plains of northern Germany, and indeed any agricultural district of that empire, produce abundant crops of potatoes and beets.

From the first, alcohol can be so easily manufactured that the processes are within the understanding and ability of any farmer. The second is used in the manufacture of beet sugar,—one of the great German industries, and the crude molasses, from a refuse product,—still contains from 40 to 50 per cent. of sugar, from which alcohol can be made. Under these circumstances and the great demand for liquid fuel for motor carriages and gas engines, alcohol for "de-naturing" came rapidly to the front as one of the most important of agricultural products, as one of the most valuable "crops" which a farmer could raise. Potatoes are chiefly raised. The potatoes are grown by the farmers and manufactured into alcohol in individual farm distilleries and in cooperative distilleries.

While England and France were somewhat behind Germany in fostering this industry—yet they both were far ahead of the United States in this matter. De-natured alcohol could be readily gotten in these countries, for industrial purposes, while the United States continued to charge a high internal revenue tax on all but wood alcohol. This prevented the use of alcohol in competition with gasoline or kerosene, and limited its use in arts and manufactures.

On June 7, 1906, however, Congress passed the "De-naturing Act," as it is called, which provided in brief that alcohol, which had been mixed with a certain proportion of de-naturing materials sufficient to prevent its use as a beverage should not be taxed.

The passage of this Act was alcohol's new day, and is destined to have a wide influence upon the agricultural pursuits of the country.

In the matter of small engines and motors alone one estimate places the farm use of these at three hundred thousand with an annual increase of one hundred thousand. This means an economical displacing of horse and muscle power in farm work almost beyond comprehension. If now the farmer can make from surplus or cheaply grown crops the very alcohol which is to furnish the cheaper fuel for his motors, he is placed in a still more independent and commanding position in the industrial race.

As an illuminant the untaxed alcohol is bound to introduce some interesting as well as novel conditions. The general estimate of the value of alcohol for lighting gives it about double the power of kerosene, a gallon of alcohol lasting as two gallons of the oil. In Germany, where the use of alcohol in lamps is most fully developed, a mantle is used. Thus in a short time it may be expected that an entirely new industry will spring up to meet the demand for the illuminating lamps embodying the latest approved form of mantle. The adapting of the gasoline motors of automobiles to alcohol fuel will in itself create a vast new manufacturing undertaking. When this is accomplished it is believed that we shall no more be troubled with the malodorous gasoline " auto " and " cycle " burners on our public streets and parkways.

De-natured alcohol is simply alcohol which has been so treated, as to spoil it for use as a beverage or medicine, and prevent its use in any manner except for industrial purposes.

De-naturing may be accomplished in many ways.

In England a mixture suitable for industrial purposes, but unfit for any other use, is made by mixing 90 per cent. of ethyl alcohol (alcohol made from grain, potatoes, beets, etc.), with 10 per cent. of methyl or " wood alcohol." Under the new law the proportion of wood alcohol is cut to five per cent.

In Canada " methylated spirits," as it is known, is composed of from 25 per cent. to 50 per cent. of wood alcohol mixed with ethyl alcohol. This proportion of wood alcohol is far more than is required in any other country.

In Germany, the de-naturing law passed in 1887 was so framed as to maintain the high revenue tax on alcohol intended for drinking, but to exempt from taxation such as should be de-naturized and used for industrial purposes. De-naturing is accomplished by mixing with the spirit a small proportion of some foreign substance, which, while not injuring its efficiency for technical uses, renders it unfit for consumption as a beverage. The de-naturing substances employed depend upon the use to which the alcohol is to be subsequently applied. They include pyridin, picolin, benzol, toluol, and xylol, wood vinegar, and several other similar products. As a result of this system Germany produced and used last year 100,000,000 gallons of de-natured spirits, as compared with 10,302,630 gallons used in 1886, the last year before the enactment of the present law.

The following are some of the other de-naturants used in Germany: Camphor, oil of turpentine, sulphuric ether, animal oil, chloroform, iodoform, ethyl bromide, benzine, castor oil, lye.

In France the standard mixture consists of:

150 liters of Ethyl alcohol,
15 liters of wood alcohol,
½ liter of heavy benzine,
1 gram. Malachite green.

An illustration of de-naturing on a large scale is given by the methods and operations of a large London establishment. On the ground floor are four large iron tanks holding about 2500 gallons each. On the next floor are casks of spirit brought under seal from the bonded warehouse. On the third floor are the wood alcohol tanks, and on the fourth floor cans of methylating materials. On the fourth floor the covers to the wood alcohol tanks were removed (these tank covers were flush with that floor) and the contents gauged and tested. The quantity to be put into the tanks on the first floor was run off through pipes connecting with the first-floor tanks and the upper tanks re-locked. Then going to the second floor, each cask of the grain spirit was gauged and tested and the tank covers, which were flush with the floor, were removed and the casks of the grain spirit were run into the tanks below. The mixture was then stirred with long-handled wooden paddles and the tank covers replaced, and the material was ready for sale free of tax. The mixture was 10 per cent. wood alcohol and 90 per cent. ethyl alcohol made from molasses, and was what is known as the ordinary methylating spirit used for manufacturing purposes only and used under bond. The completely de-natured spirit is made by adding to the foregoing three-eighths of one per cent. of benzine.

This benzine prevents re-distillation.

In the United States there are at present two general formulas for de-natured alcohol in use, either one of which may be used by any manufacturer, who can use de-natured alcohol.

The first and most common one is made up as follows:

Ethyl Alcohol	100	gallons.
Methyl "	10	"
Benzine	½	"

Where such a formula as this is required in an aqueous solution the benzine is of course thrown out, giving the solution a milky appearance. In this case the other general formula may be used.

Ethyl Alcohol	100	gallons.
Methyl "	2	"
Pyridine Bases	½	"

In addition to these two general formulas for de-natured alcohol a number of special formulas have been authorized to be used in the manufacture of certain classes of goods. In order to buy these specially de-natured alcohols it is necessary, of course, to obtain a permit first from your Collector of Internal Revenue, a simple permit to use de-natured alcohol will not suffice. Some of the special formulas are as follows:

For use in the manufacture of sulphonmethane.

Ethyl Alcohol	100	gallons.
Pyridin Bases	1	gallon.
Coal Tar Benzol	1	"

For use in the manufacture of transparent soap.

Ethyl Alcohol	100	gallons.
Methyl "	5	"
Castor Oil	1	"
36° Be. Caustic Soda Solution	½	"

For the manufacture of shellac varnishes.

Ethyl Alcohol	100	parts by volume
Methyl "	5	" " "

For the manufacture of smoking and chewing tobacco.

Ethyl Alcohol	100	gallons.
A mixture made as follows:	1	"

Aqueous Solution containing 40%
Nicotine 12 gallons
Acid Yellow Dye 0.4 lb.
Tetrazo Brilliant Blue 12 B Conct. 0.4 lb.
Water to make 100 gallons.

For the manufacture of photo-engravings.

Ethyl Alcohol	100 gallons
Sulphuric Ether	65 lbs.
Cadmium Iodide	3 "
Ammonium "	3 "

For the manufacture of fulminate of mercury.

Ethyl Alcohol	100 gallons.
Methyl "	3 "
Pyridine Bases	½ "

The next formula may be used for the following purposes:

In the manufacture of photographic dry plates.

In the manufacture of embalming fluid.

In the manufacture of heliotropin.

In the manufacture of resin of podophyllum and similar products.

In the manufacture of lacquers from soluble cotton.

In the manufacture of thermometer and barometer tubes.

Ethyl	Alcohol—	100 gallons.
Methyl	"	5 "

For use in the manufacture of photographic collodian.

Ethyl	Alcohol	100 gallons.
Sulphuric	Ether	10 lbs.
Cadmium	Iodine	10 "

For use in the manufacture of pastes and varnishes from soluble cotton.

Ethyl Alcohol	100	gallons.
Methyl "	2	"
Benzol	2	"

For use in the purification of rubber.

Ethyl Alcohol	100	gallons.
Acetone	10	"
Petroleum naptha	2	"

Petroleum naptha must have a specific gravity of not less than ·650 nor more than ·720 at 60°F.

For use in the manufacture of watches.

Ethyl Alcohol	100	gallons
Methyl "	5	"
Cyanide of Potassium	1½	lbs.
Patened Blue B	⅛	oz.

(Acid calcium, magnesium, or sodium salt of the disulpho-acids of meta-oxytetraethyldiamidotriphenyl-carbidrids.)

The methyl alcohol must have a specific gravity of not more than ·810 at 60° F.

The de-naturing mixture is best prepared by dissolving the cyanide of potassium in a small quantity of water, and then adding this solution to the alcohol, with which the methyl alcohol, containing the dissolved color, has been previously mixed.

For the manufacture of celluloid, pyralin and similar products.

Ethyl Alcohol	100	parts by volume
Methyl "	5	" " "
Camphor	7	lbs.

Alternative special de-naturant for the manufacture of celluloid, pyralin and similar products

Ethyl Alcohol	100	gallons.
Methyl "	2	"
Benzol	2	"

The strongest alcohol of commerce in the United States is usually 95 per cent. alcohol, and the price varies from $2.30 to $2.50 per gallon, showing that the greater part of the cost is due to the revenue levied by the government. The greater part of the 60,000,000 gallons of alcohol consumed in the United States is used in the manufacture of whiskey and other beverages. The revenue tax prevents the use of alcohol to any great extent in the industries of the country. The bill passed by Congress in 1906, designed to promote the use of untaxed alcohol in the arts and as fuel, took effect January 1, 1907. The first effect of free alcohol would, it was said, supplant the 12,000,000 gallons of wood alcohol which are used in the manufacture of paint, varnishes, shellacs, and other purposes. Another use that is expected of de-natured alcohol is in the manufacture of certain products, such as dyestuffs and chemicals, which can not now be manufactured commercially in this country because of the high cost of alcohol, and which are imported largely from Europe. A very rapid development of the industry of manufacturing chemicals as a result of free alcohol is looked for. In the production of alcohol there is always formed as a by-product a certain amount of fusel oil, which is very useful in manufacturing lacquers which are used on metallic susbtances, fine hardware, gas fixtures, and similar articles. The industries manufacturing these wares will undoubtedly receive a great stimulus as a result of cheaper fusel oil caused by the increased production of alcohol.

A Safe Fuel. The use of de-natured alcohol as a fuel has yet to be fully developed. Although alcohol has only about half the heating power of kerosene or gasoline, gallon for gallon, yet it has many valuable properties which may enable it to compete successfully in spite of its lower fuel value. In the first place it is very much safer. Alcohol has a tendency to simply heat the surrounding vapors and produce currents of hot gases which are not usually brought to high enough temperature to inflame articles at a distance. It can be easily diluted with water, and when it is diluted to more than one-half it ceases to be inflammable. Hence it may be readily extinguished; while burning gasoline, by floating on the water, simply spreads its flame when water is applied to it. Although alcohol has far less heating capacity than gasoline, the best experts believe that it will develop a much higher percentage of efficiency in motors than does gasoline. Since gasoline represents only about two per cent. of the petroleum which is refined, its supply is limited and its price must constantly rise, in view of the enormous demand made for it for automobiles and gasoline engines in general. This will open a new opportunity for de-natured alcohol. Industrial alcohol is now used in Germany in small portable lamps, which give it all the effects of a mantel burner heated by gas. The expense for alcohol is only about two-thirds as much per candle-power as is the cost of kerosene. Even at 25 or 30 cents a gallon, de-natured alcohol can successfully compete with kerosene as a means of lighting.

Objection has been made to the use of alcohol in automobiles and other internal-explosive engines, that it resulted in a corrosion of the metal. This is vigorously denied by the advocate of alcohol fuel and the denial is backed by proofs of the use of alcohol in German engines for a number of years without any bad results.

A recent exhibition in Germany gave a good illustration of the broad field in which de-natured alcohol may be used.

Here were shown alcohol engines of a large number of different makes, alcohol boat motors as devised for the Russian navy, and motors for threshing, grinding, wood-cutting, and other agricultural purposes.

The department of lighting apparatus included a large and varied display of lamps, chandeliers, and street and corridor lights, in which alcohol vapor is burned like gas in a hooded flame covered by a Welsbach mantle. Under such conditions alcohol vapor burns with an incandescent flame which rivals the arc light in brilliancy and requires to be shaded to adopt it to the endurance of the human eye. There has been each year a great improvement in the artistic models and finish of lamps and chandeliers for alcohol lighting. At the beginning they were simple and of rather ordinary appearance, but now they are up to the best standard of modern fixtures for gas and electricity, with which alcohol lighting is now competing with increasing success in that country.

Similarly attractive and interesting was the large display of alcohol heating stoves, which, for warming corridors, sleeping rooms, and certain other locations, are highly esteemed. They are made of japanned-iron plate in decorative forms, with concave copper reflectors, are readily portable, and, when provided with chimney connections for the escape of the gases of combustion, furnish a clean, odorless, and convenient heating apparatus.

Cooking stoves of all sizes, forms, and capacities, from the complete range, with baking and roasting ovens, broilers, etc., to the simple tea and coffee lamp, were also displayed in endless variety.

Enough has been said to give an idea of the capabilities and values of this new form of fuel,—at least, and as far as the United States is concerned.

With its advent not only will American genius perfect the machinery for its use, but the American farmer is given a new market for his crops.

Distilleries, big and little, are likely to be set up all over the country, and the time is not far distant when the farmer will be able to carry his corn to his local distillery, and either return with the money in his pocket, or with fuel for farm engines, machinery, and perchance his automobile.

When our government shall have become as far-sighted as the German government in this matter, every farmer will be able to manufacture his own de-natured spirits. The wisdom of the German system established by the law of 1887 has long ceased to be a question of debate. For every reichsmark of revenue sacrificed by exempting de-natured spirits from taxation the empire and its people have profited ten-fold by the stimulus which has been thereby given to agriculture and the industrial arts.

SPECIAL NOTICE

For the latest Government regulations regarding the manufacture of INDUSTRIAL ALCOHOL see—

REGULATIONS No. 61

Relative to the Production, Tax Payment, etc., of

INDUSTRIAL ALCOHOL

And to the Manufacture, Sale and Use of

DENATURED ALCOHOL

Under Title III of the National Prohibition Act of October 28, 1919

CHAPTER XIV.

The Free Alcohol Act of 1906, the Amendment of 1907 and Internal Revenue Regulations.

Public—No. 201.

An Act for the withdrawal from bond, tax free, of domestic alcohol when rendered unfit for beverage or liquid medicinal uses by mixture with suitable de-naturing materials.

Be it enacted by the Senate and House of Representatives of the United States of America in Congress assembled, That from and after January first, nineteen hundred and seven, domestic alcohol of such degree of proof as may be prescribed by the Commissioner of Internal Revenue, and approved by the Secretary of the Treasury, may be withdrawn from bond without the payment of internal-revenue tax, for use in the arts and industries, and for fuel, light, and power, provided said alcohol shall have been mixed in the presence and under the direction of an authorized Government officer, after withdrawal from the distillery warehouse, with methyl alcohol or other de-naturing material or materials, or admixture of the same, suitable to the use for which the alcohol is withdrawn, but which destroys its character as a beverage and renders it unfit for liquid medicinal purposes; such de-naturing to be done upon the application of any registered distillery in de-naturing bonded warehouses specially designated or set apart for de-naturing purposes only, and under conditions prescribed by the Commissioner of Internal Revenue with the approval of the Secretary of the Treasury.

The character and quantity of the said de-naturing material and the conditions upon which said alcohol may be withdrawn free of tax shall be prescribed by the Commissioner of Internal Revenue, who shall, with the approval of the Secretary of the Treasury, make all necessary regulations for carrying into effect the provisions of this Act.

Distillers, manufacturers, dealers and all other persons furnishing, handling or using alcohol withdrawn from bond under the provisions of this Act shall keep such books and records, execute such bonds and render such returns as the Commissioner of Internal Revenue, with the approval of the Secretary of the Treasury, may by regulation require. Such books and records shall be open at all times to the inspection of any internal-revenue officer or agent.

SEC. 2. That any person who withdraws alcohol free of tax under the provisions of this Act and regulations made in pursuance thereof, and who removes or conceals same, or is concerned in removing, deposting or concealing same for the purpose of preventing the same from being de-natured under governmental supervision, and any person who uses alcohol withdrawn from bond under the provision of section one of this Act for manufacturing any beverage or liquid medicinal preparation, or knowingly sells any beverage or liquid medicinal preparation made in whole or in part from such alcohol, or knowingly violates any of the provisions of this Act, or who shall recover or attempt to recover by redistillation or by any other process or means, any alcohol rendered unfit for beverage or liquid medicinal purposes under the provisions of this Act, or who knowingly uses, sells, conceals, or otherwise disposes of alcohol so recovered or redistilled, shall on conviction of each offense be fined not more than five thousand dollars, or be imprisoned not more than five years, or both, and shall, in addition, forfeit to the United States all personal property used in connection with his business, together with the buildings and lots or parcels of ground constituting the premises on which said unlawful acts are performed or permitted to be performed: *Provided*, That manufacturers employing processes in which alcohol, used free of tax under the provisions of this Act, is expressed or evaporated from the articles manufactured, shall be permitted to recover such alcohol and to have such alcohol restored to a condition suitable solely for reuse in manufacturing processes under such regulations as the Commissioner of Internal Revenue, with the approval of the Secretary of the Treasury, shall prescribe.

SEC. 3. That for the employment of such additional force of chemists, internal-revenue agents, inspectors, deputy collectors, clerks, laborers, and other assistants as the Commissioner of Internal Revenue, with the approval of the Secretary of the Treasury, may deem proper and necessary to the prompt and efficient operation and enforcement of this law, and for the purchase of locks, seals, weighing beams, gauging instruments, and for all necessary expenses incident to the proper execution of this law, the sum of two hundred and fifty thousand dollars, or so much thereof as may be required, is hereby appropriated out of any money in the Treasury not otherwise appropriated, said appropriation to be immediately available.

For a period of two years from and after the passage of this Act the force authorized by this section of this Act shall be appointed by the Commissioner of Internal Revenue, with the approval of the Secretary of the Treasury, and without compliance with the conditions prescribed by the Act entitled "An Act to regulate and improve the civil service," approved January sixteenth, eighteen hundred and eighty-three, and amendments thereof and with such compensation as the Commissioner of Internal Revenue may fix, with the approval of the Secretary of the Treasury.

SEC. 4. That the Secretary of the Treasury shall make full report to Congress at its next session of all appointments made under the provisions of this Act, and the compensation paid thereunder, and of all regulations prescribed under the provisions hereof, and shall further report what, if any, additional legislation is necessary, in his opinion, to fully safeguard the revenue and to secure a proper enforcement of this Act.

Approved June 7, 1906.

DE-NATURING REGULATIONS

Under the Act of June 7, 1906.

Under the Act quoted above, the Commissioner of Internal Revenue was empowered to make regulations whereby the law might be carried into effect.

In the first place it may be said that those who are permitted by this Act to manufacture denatured alcohol must be distillers; in other words, those who have regularly licensed and registered distilleries. This does not mean that the plant must be large or costly—as witness the numerous little "stills" to be found throughout the South; but that the still, whatever its size, must be under constant supervision, and regularly licensed to manufacture alcohol. The requirements to this end can be had from the Commissioner of Internal Revenue, Treasury Department, Washington.

Pursuant to the law regarding de-naturing, rules and regulations have been drawn up of which the following is a synopsis with extracts where deemed advisable.

De-naturing Bonded Warehouses.

"Sec. 2. The proprietor of any registered distillery may withdraw from his distillery warehouse, free of tax, alcohol of not less than 180 degrees proof or strength, to be de-natured in the manner hereinafter prescribed.

A distiller desiring to withdraw alcohol from bond for de-naturing purposes under the provisions of this act shall, at his own expense, provide a de-naturing bonded warehouse, to be situated on and constituting a part of the distillery premises. It shall be separated from the distillery and the distillery bonded warehouse and all other buildings, and no windows or doors or other openings shall be permitted in the walls of the de-naturing bonded warehouse leading into the distillery, the distillery bonded warehouse or other room or building, except as hereinafter provided. It must be constructed in the same manner as distillery bonded warehouses are now constructed, with view to the safe and secure storage of the alcohol removed thereto for de-naturing purposes and the de-naturing agents to be stored therein. It must be approved by the Commission of Internal Revenue. It shall be provided with closed mixing tanks of sufficient capacity. The capacity in wine gallons of each tank must be ascertained and marked thereon in legible letters, and each tank must be supplied with a graduated glass gauge whereon the contents will be at all times correctly indicated. All openings must be so arranged that they can be securely locked. Suitable office accommodation for the officer on duty must be provided.

SEC. 3. The de-naturing bonded warehouse shall be used for de-naturing alcohol, and for no other purpose, and nothing shall be stored or kept therein except the alcohol to be de-natured, the materials used as de-naturants, the de-natured product, and the weighing and gauging instruments and other appliances necessary in the work of de-naturing, measuring, and gauging the alcohol and de-naturing materials.

These bonded warehouses must be numbered serially in each collection district, and the words "De-naturing bonded warehouse No. —, district of —," must be in plain letters in a conspicuous place on the outside of the building.

In case the distiller's bond has been executed before the erection of such warehouse the consent of the sureties to the establishment of the de-naturing warehouse must be secured and entry duly signed made on the bond."

DE-NATURING MATERIAL ROOM.

"SEC. 4. There shall be provided within the de-naturing bonded warehouse a room to be designated as the de-naturing material room. This room is to be used alone for the storage of de-naturing materials prior to the de-naturing process. It must be perfectly secure, and must be so constructed as to render it impossible for anyone to enter during the absence of the officer in charge without the same being detected.

The ceiling, inside walls, and floor of said room must be constructed of brick, stone, or tongue-and-groove planks. If there are windows in the room the same must be secured by gratings or iron bars, and to each window must be affixed solid shutters of wood or iron, constructed in such manner that they may be securely barred and fastened on the inside. The door must be substantial, and must be so constructed that it can be securely locked and fastened.

SEC. 5. At least two sets of tanks or receptacles for storing de-naturing material must be provided, and each set of tanks must be of sufficient capacity in the aggregate to hold the de-naturing material which it is estimated the distiller will use for thirty days. A set of tanks shall consist of one or more tanks for storing methyl alcohol, and one or more tanks of smaller capacity for storing other de-naturing materials. The capacity of each tank must be ascertained and marked in legible figures on the outside.

The tanks must not be connected with each other, and must be so constructed as to leave at least 18 inches of open space between the top of the tank and ceiling, the bottom of the tank and the floor, and the sides of the tank and walls of the de-naturing material room. Each tank shall be given a number, and this number must be marked upon it. There shall be no opening at the top except such as may be necessary for dumping the de-naturing material into the tank and thoroughly plunging or mixing the same. Said opening must be covered so that it may be locked. Likewise the faucet through which the de-naturing material is drawn must be so arranged that it can be locked. Each tank must be supplied with a graduated glass gauge whereby the contents of the tank will always be shown."

Custody of De-naturing Bonded Warehouse.

"Sec. 6. The de-naturing bonded warehouse shall be under the control of the collector of the district and shall be in the joint custody of a storekeeper, storekeeper-gauger, or other designated official and the distiller.

No one shall be permitted to enter the warehouse except in the presence of said officer, and the warehouse and room shall be kept closed and the doors, exterior and interior, securely locked except when some work incidental to the process of de-naturing and storing material is being carried on. Standard Sleight locks shall be used for locking the de-naturing bonded ware-house and the de-naturing material room, and they shall be sealed in the same manner and with the same kind of seals as distillery bonded warehouses and cistern rooms are now sealed. Miller locks shall be used in securing the faucets and openings of the mixing tanks and the de-naturing material tanks.

The officer in charge of the de-naturing bonded warehouse, material room, and tanks shall carry the keys to same, and under no circumstances are said keys to be intrusted to anyone except another officer who is duly authorized to receive them."

Application for Approval of De-naturing Bonded Warehouse.

"Sec. 7. Whenever a distiller wishes to commence the business of de-naturing alcohol he must make written application to the collector of the district in which the distillery is located for the approval of a de-naturing bonded warehouse.

Such application must give the name or names of the person, firm, or corporation operating the distillery, the number of the distillery, the location of the same, the material of which the warehouse is constructed, the size of same, width, length and height, the size of the de-naturing material room therein, and the manner of its construction, the capacity in gallons of each tank to be used for de-naturing alcohol or for holding the de-naturing agents, and the material of which said tanks are constructed.

Such application must be accompanied by a diagram correctly representing the warehouse, the mixing tanks, de-naturing material room, and de-naturing material tanks, with all openings and surroundings. It must show the distillery and all the distillery bonded warehouses on the premises, with dimensions of each."

Sections 9 and 10 of the regulations deal with the examination and approval of the de-naturing warehouse and plant by the Internal Revenue officers.

De-naturing Warehouse Bond to be Given.

"Sec. 11. After receipt of notice of the approval of said warehouse the distiller may withdraw from his distillery warehouse, free of tax, alcohol of not less than 180 degrees proof or strength, and may de-nature same in said de-naturing warehouse in the manner hereinafter indicated, provided he shall first execute a bond in the form prescribed by the Commissioner of Internal Revenue, with at least two sureties, unless, under the authority contained in an act approved August 13, 1894, a corporation, duly authorized by the Attorney-General of the United States to become a surety on such bond, shall be offered as a sole surety thereon. The bond shall be for a penal sum of not less than double the tax on the alcohol it is estimated the distiller will de-nature during a period of 30 days, and in no case is the distiller to withdraw from bond for de-naturing purposes and have in his de-naturing warehouse in process of de-naturation a quantity of alcohol the tax upon which is in excess of the penal sum of the bond.

SEC. 12. If at any time, it should develop that the de-naturing warehouse bond is insufficient the distiller must give additional bond.

SEC. 13. The bond herein provided for must be executed before the distiller can withdraw from distillery bonded warehouse, free of tax, alcohol to be de-natured, and if he desires to continue in the business of de-naturing alcohol, said bond must be renewed on the first day of May of each year or before any alcohol is withdrawn from bond for de-naturing purposes. It must be executed in duplicate in accordance with instructions printed thereon. One copy is to be retained by the collector and one copy is to be transmitted to the Commissioner of Internal Revenue."

CONDITIONS UNDER WHICH ALCOHOL IS WITHDRAWN.

"SEC. 15. Not less than three hundred (300) wine gallons of alcohol can be withdrawn at one time for de-naturing purposes.

When a distiller, who is a producer of alcohol of not less than 180 degrees proof and who has given the de-naturing warehouse bond as aforesaid desires to remove alcohol from the distillery bonded warehouse for the purpose of de-naturing, he will himself, or by his duly authorized agent, file with the collector of internal revenue of the district in which the distillery is located, notice to that effect."

Upon the receipt of this notice (the form for

which is given in the Regulations) the collector for the district will order a gauger to inspect the alcohol so withdrawn, and to gauge the same, and to make report; and directions are given to the official "storekeeper" to permit the transferral of the spirits to the de-naturing warehouse.

Spirits Transferred to be Marked.

"Upon receipt of the permit by the storekeeper the packages of distilled spirits described in notice of intention to withdraw may be withdrawn from distillery bonded warehouse without the payment of the tax, and may be transferred to the de-naturing bonded warehouse on the distillery premises; but before the removal of said spirits from the distillery bonded warehouse, the gauger, in addition to marking, cutting, and branding the marks usually required on withdrawal of spirits from warehouse, will legibly and durably mark on the head of each package, in letters and figures not less than one-half an inch in length, the number of *proof* gallons then ascertained, the date of the collector's permit, the object for which the spirits were withdrawn, and his name, title, and district.

Such additional marks may be as follows:

Withdrawn under permit issued Jan'y. 10, 1907
For De-naturing Purposes
Proof gallons, 84
William Williams, U. S. Gauger,
5th Dist. *Ky*."

Spirits Transferred to De-naturing Bonded Warehouse.

"Sec. 20. When the packages of spirits are marked and branded in the manner above indicated they shall at once, in the presence and under the supervision of the storekeeper, be transferred to the de-naturing bonded warehouse."

Record of Spirits Received in De-naturing Bonded Warehouse.

"Sec. 21. The officer in charge of the de-naturing bonded warehouse shall keep a record of the spirits received in said de-naturing bonded warehouse from the distillery bonded warehouse and the spirits delivered to the distiller for de-naturing purposes.

Upon the *debit* side of said record, in columns prepared for the purpose, there shall be entered the date when any distilled spirits were received in de-naturing bonded warehouse, the date of the collector's permit, the date of withdrawal from distillery bonded warehouse, the number of packages received, the serial numbers of the packages, the serial numbers of the distillery warehouse stamps, and the wine and proof gallons.

Upon the *credit* side of said record shall be entered the date when any spirits were delivered to the distiller for de-naturing purposes, the date of the collector's permit for withdrawal, the date of withdrawal from distillery bonded warehouse, the number of packages so delivered, the serial numbers of the packages, the serial numbers of the distillery warehouse stamps, and the wine and proof gallons.

Immediately upon the receipt of any distilled spirits in the de-naturing bonded warehouse, and on the same day upon which they are received, the officer must enter said spirits in said record.

Likewise, on the same date upon which any spirits are delivered to the distiller for de-naturing purposes, said spirits must be entered on said record.

SEC. 22. A balance must be struck in the record described in above section at the end of the month showing the number of packages and quantity in wine and proof gallons of spirits on hand in packages on the first day of the month, the number of packages and quantity in wine and proof gallons received during the month, the number of packages and quantity in wine and proof gallons delivered to the distiller during the month, and the balance on hand in packages and wine and proof gallons at the close of the month."

Sections 23 to 25 of the Rules relate to the duties of the Internal Revenue officers in making reports and returns.

DE-NATURING AGENTS. COMPLETELY DE-NATURED ALCOHOL.

" SEC. 26. Unless otherwise specially provided, the agents used for de-naturign alcohol withdrawn from bond for de-naturing purposes shall consist of methyl alcohol and benzine in the following proportions: To every 100 parts by volume of ethyl alcohol of the desired proof (not less than 180°) there shall be added 10 parts by volume of approved methyl alcohol and one-half of one part by volume of approved benzine; for example, to every 100 gallons of ethyl alcohol (of not less than 180 degrees proof) there shall be added 10 gallons of approved methyl alcohol and one-half gallon of approved benzine. Alcohol thus de-natured shall be classed as completely de-natured alcohol.

Methyl alcohol and benzine intended for use as de-naturants must be submitted for chemical test and must conform to the specifications which shall be hereafter duly prescribed."

De-naturants Deposited in Warehouse.

"Sec. 27. As the distiller's business demands, he may bring into the de-naturing bonded warehouse, in such receptacles as he may wish, any authorized de-naturant. Such de-naturants shall at once be deposited in the material room; thereafter they shall be in the custody and under the control of the officer in charge of the warehouse. Before any de-naturant is used it must be dumped into the appropriate tank and after the contents have been thoroughly mixed, a sample of one pint taken therefrom. This sample must be forwarded to the proper officer for analysis. The officer will then securely close and seal the tank.

No part of the contents of the tank can be used until the sample has been officially tested and approved, and report of such test made to the officer in charge of the warehouse.

If the sample is approved the contents of the tank shall upon the receipt of the report, become an approved de-naturant and the officer shall at once remove the seals and place the tank under Government locks.

If the sample does not meet the requirements of the specifications, the officer shall, upon the receipt of the report of non-approval, permit the distiller, provided he desires, to treat or manipulate the proposed de-naturant so as to render it a competent de-naturant. In such case another sample must be submitted for approval. If the distiller does not desire to further treat the de-naturant the officer shall require him immediately to remove the contents of the tank from the premises."

Record of De-naturants Received.

" Sec. 28. The officer shall keep a de-naturing material room record. This record shall show all material entered into and removed from the de-naturing material room.

There shall be proper columns on the *debit* side in which are to be entered the date when any material is received, the name and residence of the person from whom received, the kind of material, the quantity in wine gallons, and, if methyl alcohol, in proof gallons, the date upon which the material was dumped into the tank, the number of the tank, the date upon which sample was forwarded, and the number of the sample, and the result of the official test.

On the *credit* side of said record shall be entered in proper columns the date upon which any material was removed from the de-naturing material room for de-naturing purposes, the kind of material, the number of the tank from which taken, the number of the sample representing the tank and sent for official test, the number of wine gallons, and, if methyl alcohol, the number of proof gallons."

Monthly Returns of De-naturants Received.

"Sec. 29. A balance shall be struck in this record at the end of each month whereby shall be shown the quantity of material of each kind on hand in the de-naturing material room on the first day of the month, the quantity received during the month, the quantity rejected and removed from the premises during the month, and the quantity delivered to the distiller for de-naturing purposes during the month, and the quantity on hand at the end of the month.

The officer shall, at the end of each month, prepare in duplicate, sign, and forward to the collector of internal revenue a report which shall be a transcript of said record."

DISTILLER TO *KEEP* RECORD OF DE-NATURANTS.

"SEC. 30. The distiller shall also keep a record, in which he shall enter the date upon which he deposits any material in the tanks of the de-naturing material room, the name and address of the person from whom said material was received, and the kind and quantity of the material so deposited; also he shall enter in said record the date upon which he receives any material from the de-naturing material room, the kind and quantity of such material so received, and the disposition made of same."

NOTICE OF INTENTION TO DE-NATURE SPIRITS.

"SEC. 31. The distiller shall, before dumping any spirits or de-naturants into the mixing tank, give notice to the officer in charge of the de-naturing warehouse in proper form in duplicate, and enter in the proper place thereon (in the case of distilled spirits) and in the proper column the number of the packages, the serial numbers of same, the serial number of the warehouse stamps, the contents in wine and proof gallons and the proof as shown by the marks, the date of the withdrawal gauge, and by whom gauged.

In case of de-naturing agents he shall enter in the proper place and in the proper columns the number of gallons, the kind of material, and the number of the de-naturing material tank from which same is to be drawn.

The contents of the several packages of alcohol, as shown by the withdrawal gauge, shall be accepted as the contents of said packages when dumped for de-naturing purposes unless it should appear from a special showing made by the distiller that there has been an accidental loss since withdrawal from distillery bonded warehouse.

Upon receipt of this notice the officer in charge of the de-naturing warehouse shall, in case of the packages of alcohol, inspect same carefully to ascertain whether or not they are the packages described in the distiller's notice. He will then cut out that portion of the warehouse stamp upon which is shown the serial number of the stamp, the name of the distiller, the proof gallons, and the serial number of the package. These slips must be securely fastened to the form whereon the gauging is reported and sent by the officer with his return to the collector."

Transfer of De-naturants to Mixing Tanks.

"Sec. 32. The distiller, unless pipes are used, as herein provided, shall provide suitable gauged receptacles, metal drums being preferred, with which to transfer the de-naturing agents from the material tanks to the mixing tanks. These receptacles must be numbered serially and the number, the capacity in gallons and fractions of a gallon, the name of the distiller, and the number of the de-naturing bonded warehouse marked thereon in durable letters and figures. They shall be used for transferring de-naturing material from the material tanks to the mixing tanks and for no other purpose. The distiller must also provide suitable approved sealed measures of smaller capacity. The gauged receptacles are to be used where the quantity to be transferred amounts to as much as the capacity of the smallest gauged receptacle in the warehouse. The measures are to be used only when the quantity of material to be transferred is less than the capacity of the smallest gauged receptacle.

SEC. 33. The distiller may provide metal pipes connecting the material tanks and the mixing tanks and the de-naturant may be transferred to the mixing tanks through these pipes. Such pipes must be supplied with valves, cocks, or faucets, other proper means of controlling the flow of the liquid, and such valves, cocks, or faucets must be so arranged that they can be securely locked, and the locks attached thereto must be kept fastened; the keys to be retained by the officer in charge, except when the de-naturing material is being transferred to the mixing tanks.

In the event pipes are used as above provided, the glass gauges affixed to the material tanks must be so graduated that tenths of a gallon will be indicated.

Before any material is transferred from a material tank to a mixing tank the officer must note the contents of the material tank as indicated by the glass gauge. He will then permit the de-naturant to flow into the mixing tank until the exact quantity necessary to de-nature the alcohol, as provided by the regulations, has been transferred. This he will ascertain by reading the gauge on the material tank before the liquid has begun to flow and after the flow has been stopped. He should verify the quantity transferred by reading the gauge on the mixing tank before and after the transfer.

SEC. 34. The distiller must provide all scales, weighing beams, and other appliances necessary for transferring the de-naturing materials gauging or handling the alcohol, or testing any of the measures, receptacles or gauges used in the warehouse, and also a sufficient number of competent employees for the work.

CONTENTS OF MIXING TANK TO BE PLUNGED.

"SEC. 35. The exact quantity of distilled spirits contained in the packages covered by the distiller's notice having been ascertained by the officer and the spirits having been dumped into the mixing tank, and the quantities of the several de-naturants prescribed by the regulations having been ascertained by calculation and added as above provided to the alcohol in the mixing tank to be thoroughly and completely plunged and mixed by the distiller or his employees."

DRAWING OFF AND GAUGING DE-NATURED PRODUCT.

"SEC. 37. The distiller may from time to time as he wishes, in the presence of the officer, draw off from the tank or tanks the de-natured product in quantities of not less than 50 gallons at one time, and the same must at once be gauged, stamped, and branded by the officer and removed from the premises by the distiller."

KIND AND CAPACITY OF PACKAGES USED.

"SEC. 38. He may use packages of a capacity of not less than five gallons or not more than one hundred and thirty-five (135) gallons, and each package must be filled to its full capacity, such wantage being allowed as may be necessary for expansion.

All packages used to contain completely de-natured alcohol must be painted a *light green*, and in no case is a package of any other color to be used."

ALCOHOL TO BE IMMEDIATELY DE-NATURED.

"SEC. 39. No alcohol withdrawn from distillery warehouse for de-naturing purposes shall be permitted to remain in the de-naturing bonded warehouse until after the close of business on the second day after the said alcohol is withdrawn, but all alcohol so withdrawn must be transferred, dumped, and de-natured before the close of business on said second day."

APPLICATION FOR GAUGE OF DE-NATURED ALCOHOL.

"SEC. 40. When the process of de-naturing has been completed and the distiller desires to have the de-natured alcohol drawn off into packages and gauged, he shall prepare a request for such gauge on the proper form. The request shall state as accurately as practicable the number of packages to be drawn off and the number of wine and proof gallons contents thereof.

This notice shall be directed to the collector of internal revenue, but shall be handed to the officer on duty at the de-naturing bonded warehouse.

SEC. 41. If the officer shall find upon examination of the proper record that there should be on hand the quantity of de-natured alcohol covered by said notice, he shall proceed to gauge and stamp the several packages of de-natured alcohol in the manner herein prescribed, and shall make report thereof on the proper form.

In no case will the officer gauge and stamp de-natured alcohol the total quantity in wine gallons of which taken together with any remnant that may be left in the de-naturing tank exceeds in wine gallons the sum of the quantity of distilled spirits and de-naturants dumped on that day and any remnant brought over from previous day."

HOW DE-NATURED ALCOHOL SHALL BE GAUGED.

"SEC. 42. The gauging of de-natured alcohol shall, where it is practicable, be by weight. The officer shall ascertain the tare by actually weighing each package when empty. Then, after each package has been filled in his presence, he shall ascertain the gross weight, and, by applying the tare, the net weight.

He shall then ascertain the proof in the usual manner, and by applying the proof to the wine gallons content the proof gallons shall be ascertained.

The regulations relating to the gauging of rectified spirits, so far as they apply to apparent proof and apparent proof gallons, shall apply to de-natured spirits. Where it is for any reason not practicable to gauge de-natured alcohol by weight, using the tables that apply in the case of the gauging of distilled spirits, the gauging shall be by rod."

Sections 43 to 45 provide for the returns to be made by the Government officials, and the proper marking of the packages containing de-natured alcohol; and Sections 46 to 48 lay down the form of the Government stamps and their use.

Section 49 places the mixing tank absolutely in the control of the warehouse officer, and requires if he leaves the warehouse he must close and lock the same.

Section 50 deals with records to be kept by warehouse officer.

De-natured Alcohol to be Removed from Warehouse.

"Sec. 51. Not later than the close of business on the day following that upon which the work of drawing off and gauging the de-natured spirits is completed, the distiller must remove said de-natured alcohol from the de-naturing bonded warehouse. He may either remove the alcohol to a building off the distillery premises, where he can dispose of it as the demands of the trade require, or he may dispose of it in stamped packages direct to the trade from the de-naturing bonded warehouse."

Sections 52 and 53 relate to records to be kept by the distiller showing de-natured alcohol received and disposed of by him, and the parties to whom the same was sold or delivered. Sections 54 to 57 cover reports and records to be made by officers and collector.

Part II of the Regulations relates to dealers in de-natured alcohol, and manufacturers using the same.

"SEC. 58. Alcohol de-natured by use of methyl alcohol and benzine as provided in section 26 of these regulations is to be classed as *completely de-natured alcohol*. Alcohol de-natured in any other manner will be classed as *specially de-natured alcohol*."

De-natured Alcohol not to be Stored on Certain Premises, and not to be Used for Certain Purposes.

"SEC. 59. Neither completely nor specially de-natured alcohol shall be kept or stored on the premises of the following classes of persons, to wit: dealers in wines, fermented liquors or distilled spirits, rectifiers of spirits, manufacturers of and dealers in beverages of any kind, manufacturers of liquid medicinal preparations, or distillers (except as to such de-natured alcohol in stamped packages as is manufactured by themselves), manufacturers of vinegar by the vaporizing process and the use of a still and mash, wort, or wash, and persons who, in the course of business, have or keep distilled spirits, wines, or malt liquors, or other beverages stored on their premises. *Provided*, That druggists are exempt from the above provisions."

Can Not be Used in Manufacturing Beverages, etc.

"Sec. 60. Anyone using de-natured alcohol for the manufacture of any beverage or liquid medicinal preparation, or who knowingly sells any beverage or liquid medicinal preparation made in whole or in part from such alcohol, becomes subject to the penalites prescribed in section 2 of the Act of June 7, 1906."

Under the language of this law it is held that de-natured alcohol can not be used in the preparation of any article to be used as a component part in the preparation of any beverage or liquid medicinal preparation.

A person, firm, or corporation desiring to sell de-natured alcohol, must make application, in proper form, to the district collector on or before the first of July each year, and if the provisions of the law have been violated the permit may be withdrawn (Sections 61 to 65).

Sections 66 to 71 relate to the keeping of records by collector, and wholesale and retail dealers.

RETAIL DEALERS TO KEEP RECORD.

"SEC. 72. Retail dealers in de-natured alcohol shall keep a record, in which they shall enter the date upon which they receive any package or packages of de-natured alcohol, the person from whom received, the serial numbers of the packages, the serial numbers of the de-natured alcohol stamps the wine and proof gallons, and the date upon which packages are opened for retail.

The transcript for each month's business as shown by this record must be prepared, signed, and sworn to and forwarded to the collector of internal revenue of the district in which the dealer is located before the 10th of the following month. This transcript must be signed and sworn to by the dealer himself or by his duly authorized agent."

LABELS TO BE PLACED ON RETAIL PACKAGES.

"SEC. 73. Retail dealers in de-natured alcohol must provide themselves with labels upon which the words "De-Natured Alcohol" have been printed in plain, legible letters. The printing shall be red on white. A label of this character must be affixed by the dealer to the container, whatever it may be, in the case of each sale of de-natured alcohol made by him."

Stamps to be Destroyed when Package is Empty.

"Sec. 74. As soon as the stamped packages of de-natured alcohol are empty the dealer or manufacturer, as the case may be, must thoroughly obliterate and completely destroy all marks, stamps, and brands on the packages.

The stamps shall under no circumstances be re-used, and the packages shall not be refilled until *all* the marks, stamps, and brands shall have been removed and destroyed."

Manufacturers Using Completely De-natured Alcohol to Secure Permit.

"Sec. 75. Manufacturers desiring to use completely de-natured alcohol, such as is put upon the market for sale generally, may use such alcohol in their business subject to the following restrictions:

A manufacturer using less than an average of 50 gallons of de-natured alcohol per month will not be required to secure permit from the collector or to keep records or make returns showing the alcohol received and used.

Manufacturers who use as much as 50 gallons of completely de-natured alcohol a month must procure such alcohol in stamped packages, and before beginning business the manufacturer must make application to the collector of the proper district for permit, in which application he will state the exact location of his place of business, describing the lot or tract of land upon which the plant is located, and must keep the alcohol in a locked room until used.

"SEC. 79. As the agents adapted to and adopted for use in complete de-naturation render the alcohol de-natured unfit for use in many industries in which ethyl alcohol, withdrawn free of tax, can be profitably employed, therefore in order to give full scope to the operation of the law, special de-naturants will be authorized when absolutely necessary. Yet the strictest surveillance must be exercised in the handling of alcohol incompletely or specially de-natured."

FORMULA FOR SPECIAL DE-NATURANTS TO BE SUBMITTED TO THE COMMISSIONER.

"SEC. 80. The Commissioner of Internal Revenue will consider any formula for special de-naturation that may be submitted by any manufacturer in any art or industry and will determine (1) whether or not the manufacture in which it is proposed to use the alcohol belongs to a class in which tax-free alcohol withdrawn under the provisions of this act can be used. (2) whether or not it is practicable to permit the use of the proposed de-naturant and at the same time properly safeguard the revenue. But one special de-naturant will be authorized for the same class of industries, unless it shall be shown that there is good reason for additional special de-naturants."

The Commissioner will announce from time to

time the formulas of de-naturants that will be permitted in the several classes of industries in which tax-free alcohol can be used.

The specially or incompletely de-natured alcohol can only be used by special permission, for which the manufacturer must apply, at the same time giving full details as to business, plant, premises, the special de-naturants desired to be used, and the reason therefor, etc. (Section 81).

Section 82 recites the necessary requirements as to storerooms, etc., and Sections 83 to 87 relate to the form of application and the inspection of the plant. Section 88 recites the form of bond necessary to be given by the manufacturer, and Sections 89 to 104 relate to the general requirements as to records, books, affidavits, etc.

Sections 105 and 106 rule that the alcohol must be used just as received, and as called for in the permit, and that a manufacturer quitting business may dispose of his specially de-natured alcohol to other manufacturers.

Provisions Applicable to Manufacturers Using Either Specially or Generally De-Natured Alcohol.

"Sec. 107. Under no circumstances will de-naturers, manufacturers, or dealers, or any other persons, in any manner treat either specially or completely de-natured alcohol by adding anything to it or taking anything from it until it is ready for the use for which it is to be employed. It must go into manufacture or consumption in exactly the same condition that it was when it left the de-naturer. Diluting completely de-natured alcohol will be held to be such manipulation as is forbidden by law.

"SEC. 108. Manufacturers using either specially or completely de-natured alcohol must store it in the storeroom set apart for that purpose, the place for deposit named in the bond and application, and nowhere else. Likewise they must deposit recovered alcohol in said storeroom as fast as it is recovered. It will be held to be a breach of the bond and a violation of the law if any alcohol of any kind, character, or description should be found stored at any other place on the premises."

The question of special de-naturants is one of great importance to the manufacturer, and should be carefully studied. The distiller who succeeds on a large scale will be he who is most expert in preparing alcohol specially de-natured to suit the requirements of the various arts. Germany has done most in this line, and the German practice should be carefully studied.

Parts IV and V of the Rules relate to that portion of the De-Naturing Act, referred to in Section 2 thereof—the recovering, restoring and re-denaturing of alcohol used by manufacturers employing processes in which the formerly de-natured spirits is expressed, or evaporated. This not being within the plan of this book, the rules relating thereto are not quoted.

Those desirous of acquiring full information as to the rules regulating the operation of distilleries for the manufacture of alcohol and de-natured spirits can procure the same by applying either to the collectors of Internal Revenue for their respective districts or to the Commissioner of Internal Revenue, Washington, D. C.

PROPOSED CHANGES IN THE DE-NATURING ACT.

The De-naturing Act as passed and the regulations thereunder are undoubtedly too complicated in their character to remain very long in the Statute Books. There has already arisen a cry for simpler regulations which shall place the manufacture of de-natured alcohol on a plane with the practice in Germany, France and other countries which have carried the manufacture and use of alcohol for industrial purposes to a very high plane. Both in England and America the Excise and Internal Revenue regulations have been of very troublesome character, and the production of spirits has been so carefully guarded, watched and checked that the distiller aside from the high tax he has had to pay has been greatly hampered. In Germany and France, however, things are different. There the manufacture of Industrial Alcohol from farm products has been encouraged and as a consequence the regulations are of very much simpler character. In Germany the number of agricultural or co-operative stills is very large and these stills are practically free from the constant supervision of internal revenue officials.

Until the wash passes into the still there is practically no Governmental supervision except as to the proper gauging of the vats and to the proper sealing of all joints or pipes leading from the vats to the still. From that point onward, however, to the final receiver every vessel is locked and sealed and no access to the spirit can be obtained by the distiller. The quantity of spirit distilled and its quality is ascertained by the Revenue Officer from this final receiver and on this spirit so found is computed the vat tax and the distillery tax which have to be paid by the distiller. There are none of the cumbersome regulations regarding the warehouses, storehouses, storekeepers, etc., which are found in our own revenue laws. To provide security against abstraction of wash in the fermenting tanks, reliance is placed upon frequent but uncertain visitations.

There is no question but that in the fulness of time our own laws and regulations will be very much simplified for all industrial plants. An attempt has been made to so simplify the laws by Act of Congress No. 230, approved March 2, 1907 and taking effect on September 1, 1907, the text of which is appended, and undoubtedly other acts will follow as the country becomes more and more sensible of the benefits to be derived from free industrial alcohol. The text of the act is as follows:

[PUBLIC—NO. 230.]

An Act to amend an Act entitled "An Act for the withdrawal from bond tax free of domestic alchohol when rendered unfit for beverage or liquid medicinal uses by mixture with suitable denaturing materials," approved June seventh, nineteen hundred and six.

Be it enacted by the Senate and House of Representatives of the United States of America in Congress Assembled, That notwithstanding anything contained in the Act entitled "An Act for the withdrawal from bond tax free of domestic alcohol when rendered unfit for beverage or liquid medicinal uses by mixture with suitable de-naturing materials," approved June seventh, nineteen hundred and six, domestic alcohol when suitably denatured may be withdrawn from bond without the payment of internal-revenue tax and used in the manufacture of ether and chloroform and other definite chemical substances where said alcohol is changed into some other chemical substance and does not appear in the finished product as alcohol: *Provided*, That rum of not less than one hundred and fifty degrees proof, may be withdrawn, for de-naturation only, in accordance with the provisions of said Act of June seventh, nineteen hundred and six, and in accordance with the provisions of this Act.

SEC. 2. That the Commissioner of Internal Revenue, with the approval of the Secretary of the Treasury, may authorize the establishment of central de-naturing bonded warehouses, other than those at distilleries, to which alcohol of the required proof may be transferred from distilleries or distillery bonded warehouses without the payment of internal-revenue tax, and in which such alcohol may be stored and de-natured. The establishment, operation, and custody of such warehouses shall be under such regulations and upon the execution of such bonds as the Commissioner of Internal Revenue, with the approval of the Secretary of the Treasury, may prescribe.

Sec. 3. That alcohol of the required proof may be drawn off, for de-naturation only, from receiving cisterns in the cistern room of any distillery for transfer by pipes direct to any de-naturing bonded warehouse on the distillery premises or to closed metal storage tanks situated in the distillery bonded warehouse, or from such storage tanks to any denaturing bonded warehouse on the distillery premises, and de-natured alcohol may also be transported from the de-naturing bonded warehouse, in such manner and by means of such packages, tanks or tank cars, and on the execution of such bonds, and under such regulations as the Commissioner of Internal Revenue, with the approval of the Secretary of the Treasury, may prescribe. And further, alcohol to be de-natured may be withdrawn without the payment of internal-revenue tax from the distillery bonded warehouse for shipment to central de-naturing plants in such packages, tanks and tank cars, under such regulations, and on the execution of such bonds as may be prescribed by the Commissioner of Internal Revenue, with the approval of the Secretary of the Treasury.

SEC. 4. That at distilleries producing alcohol from any substance what ever, for de-naturation only, and having a daily spirit-producing capacity of not exceeding one hundred proof gallons, the use of cisterns or tanks of such size and construction as may be deemed expedient may be permitted in lieu of distillery bonded warehouses, and the production, storage, the manner and process of denaturing on the distillery premises the alcohol produced, and transportation of such alcohol, and the operation of such distilleries shall be upon the execution of such bonds and under such regulations as the Commissioner of Internal Revenue, with the approval of the Secretary of the Treasury, may prescribe, and such distilleries may by such regulations be exempted from such provisions of the existing laws relating to distilleries as may be deemed expedient by said officials.

SEC. 5. That the provisions of this Act shall take effect on September first, nineteen hundred and seven.

Approved, March 2, 1907.

INDEX.

Spotted something that isn't quite right?

Let us know and we'll fix it for future reprints. And as a thank you for your diligence, we'll send you a free Sothis book of your choice.

publisher@sothis.press

This page intentionally left blank

Except of course for that line above. And this one.
And whilst we have your attention, perhaps you'd
like to follow us on Instagram @sothis.press. Go on.